Metric Mechanical Mathematics

ANSWER BOOK

by A. J. Raven and S. M. Ault

Mathematics for Everyday Life
Books 1 to 5

Metric Mechanical Mathematics

ANSWER BOOK

A. J. RAVEN

Mathematics Master, King Ethelbert County Secondary School, Birchington

Heinemann Educational Books
London

Heinemann Educational Books Ltd

LONDON EDINBURGH MELBOURNE AUCKLAND TORONTO
SINGAPORE HONG KONG KUALA LUMPUR
IBADAN NAIROBI JOHANNESBURG
NEW DELHI

ISBN 0 435 50807 5
First published 1972
Reprinted 1972

Published by Heinemann Educational Books Ltd
48 Charles Street, London W1X 8AH

Printed in Great Britain by
The Whitefriars Press Ltd., London and Tonbridge

Answers

EXERCISE 1 (page 1)

1. 357	**2.** 1408	**3.** 1681	**4.** 3648
5. 13 307	**6.** 18 408	**7.** 18 705	**8.** 19 499
9. 17 536	**10.** 19 850	**11.** 16 177	**12.** 29 723
13. 34 798	**14.** 32 972	**15.** 33 803	**16.** 306 566
17. 143 736	**18.** 228 605	**19.** 294 405	**20.** 316 066
21. 259 325	**22.** 276 527	**23.** 369 390	**24.** 261 211
25. 197 780	**26.** 328 320	**27.** 282 719	**28.** 224 258
29. 335 308	**30.** 313 389	**31.** 2048	**32.** 2088
33. 17 922	**34.** 5044	**35.** 3931	**36.** 16 113
37. 21 144	**38.** 16 700	**39.** 59 119	**40.** 43 274
41. 123 566	**42.** 181 986	**43.** 129 414	**44.** 220 270
45. 113 873	**46.** 101 572	**47.** 35 294	**48.** 136 511
49. 139 794	**50.** 118 809		

EXERCISE 2 (page 2)

1. 572	**2.** 522	**3.** 636	**4.** 3745
5. 2357	**6.** 2609	**7.** 5456	**8.** 6899
9. 2727	**10.** 3979	**11.** 4579	**12.** 360
13. 939	**14.** 3989	**15.** 3671	**16.** 561
17. 1128	**18.** 79 164	**19.** 68 292	**20.** 75 501
21. 2051	**22.** 41 187	**23.** 11 105	**24.** 17 477
25. 57 911	**26.** 38 863	**27.** 8430	**28.** 3210
29. 3206	**30.** 3208	**31.** 7006	**32.** 8327
33. 725	**34.** 763	**35.** 10 622	**36.** 156
37. 252	**38.** 536	**39.** 519	**40.** 562
41. 409	**42.** 539	**43.** 639	**44.** 489
45. 786	**46.** 2192	**47.** 3091	**48.** 1979
49. 2878	**50.** 6883	**51.** 3587	**52.** 1079
53. 5887	**54.** 5989	**55.** 5519	**56.** 5653
57. 4521	**58.** 9271	**59.** 8652	**60.** 7733
61. 57 473	**62.** 41 007	**63.** 40 278	**64.** 37 876
65. 21 871	**66.** 66 877	**67.** 18 074	**68.** 53 883
69. 50 677	**70.** 8864		

EXERCISE 3 (page 3)

1. 640	2. 740	3. 1880	4. 6230
5. 4740	6. 7560	7. 272	8. 294
9. 544	10. 627	11. 1053	12. 2597
13. 4864	14. 3724	15. 7812	16. 3420
17. 3591	18. 2523	19. 1872	20. 6460
21. 3953	22. 4464	23. 6764	24. 6370
25. 5056	26. 2773	27. 8526	28. 4410
29. 8140	30. 12 690	31. 7192	32. 19 352
33. 27 072	34. 13 482	35. 6048	36. 27 768
37. 17 892	38. 26 296	39. 42 228	40. 28 188
41. 58 167	42. 46 718	43. 12 691	44. 17 496
45. 10 266	46. 44 744	47. 30 409	48. 9131
49. 17 316	50. 56 192	51. 37 387	52. 53 235
53. 20 748	54. 46 342	55. 40 752	56. 74 497
57. 36 666	58. 59 904	59. 82 302	60. 51 903
61. 82 353	62. 60 352	63. 23 814	64. 18 522
65. 20 748	66. 71 586	67. 55 024	68. 70 626
69. 56 615	70. 41 025	71. 45 312	72. 84 318
73. 39 903	74. 23 436	75. 37 142	76. 20 000
77. 41 000	78. 66 456	79. 133 076	80. 46 248
81. 68 694	82. 55 208	83. 53 505	84. 171 41
85. 190 944	86. 109 915	87. 390 011	88. 158 23
89. 289 209	90. 163 048	91. 545 421	92. 396 03
93. 576 384	94. 585 216	95. 804 003	96. 602 89
97. 761 816	98. 789 342	99. 850 406	100. 881 34

EXERCISE 4 (page 4)

1. 211 rem.2	2. 184 rem.3	3. 79 rem.4
4. 127 rem.5	5. 118 rem.2	6. 94 rem.3
7. 256 rem.1	8. 2121	9. 1157 rem.4
10. 563 rem.6	11. 1062 rem.1	12. 552 rem.8
13. 1498 rem.3	14. 3207	15. 995
16. 1415 rem.1	17. 1795	18. 622
19. 1062 rem.1	20. 6928 rem.1	21. 4330 rem.6
22. 10 653 rem.3	23. 14 995 rem.1	24. 12 123 rem.2
25. 3307 rem.1	26. 217 rem.10	27. 207 rem.10
28. 239 rem.8	29. 270 rem.6	30. 745 rem.9
31. 415 rem.7	32. 777 rem.5	33. 448 rem.9

34. 328 rem.8	35. 566 rem.7	36. 580 rem.14
37. 480 rem.3	38. 189 rem.7	39. 216 rem.10
40. 349 rem.9	41. 320 rem.13	42. 257 rem.11
43. 238 rem.21	44. 175 rem.6	45. 95 rem.36
46. 136 rem.47	47. 125 rem.36	48. 7725 rem.5
49. 1489 rem.4	50. 1303 rem.1	51. 1085 rem.3
52. 1620 rem.3	53. 1622 rem.9	54. 1156 rem.13
55. 1326 rem.3	56. 2201 rem.9	57. 2190 rem.1
58. 2949 rem.4	59. 2137 rem.12	60. 1113 rem.32
61. 790 rem.22	62. 168 rem.45	63. 282 rem.27
64. 639 rem.13	65. 4934 rem.16	66. 1773 rem.20
67. 1970 rem.12	68. 846	69. 1149 rem.30
70. 550 rem.40	71. 820 rem.61	72. 1264 rem.31
73. 1794 rem.3	74. 1241 rem.4	75. 1109 rem.10
76. 819 rem.60	77. 464 rem.21	78. 835 rem.1
79. 2050 rem.21	80. 1297 rem.22	81. 1181 rem.6
82. 724 rem.21	83. 1055 rem.55	84. 117 rem.63
85. 209 rem.14	86. 2034 rem.1	87. 2185 rem.99
88. 3701 rem.20	89. 9144 rem.68	90. 4270 rem.57
91. 2439 rem.65	92. 1375 rem.62	93. 6899 rem.46
94. 3853 rem.33	95. 3414 rem.39	96. 852 rem.281
97. 1791 rem.363	98. 3216 rem.184	99. 4058 rem.10
100. 4505 rem.144		

EXERCISE 5 (page 5)

Test 1

1. 357	2. 1840	3. 108 781	4. 21 086
5. 91 976	6. 296	7. 1277	8. 891
9. 1002	10. 129	11. 7633	12. 6764
13. 18 723	14. 50 184	15. 153 032	16. 66 456
17. 232 764	18. 531 rem.3	19. 7480 rem.2	
20. 1262 rem.13	21. 4015 rem.7	22. 694 rem.23	
23. 2418 rem.19	24. 811 rem.23	25. 6899 rem.68	

Test 2

1. 1022	2. 10 292	3. 1143	4. 21 800
5. 10 841	6. 254 337	7. 903 380	8. 419
9. 274	10. 5738	11. 3871	12. 6219

13. 45 017	14. 3257	15. 1073	16. 7812
17. 67 071	18. 34 983	19. 28 782	20. 365 67(
21. 761 rem.2	22. 401 rem.4	23. 308 rem.17	
24. 541 rem.15	25. 7216 rem.43		

Test 3

1. 2265	2. 22 650	3. 6830	4. 50 942
5. 108 630	6. 904 945	7. 492	8. 571
9. 6489	10. 81 329	11. 18 633	12. 11 239
13. 891	14. 2793	15. 8633	16. 5781
17. 148 106	18. 404 124	19. 438 213	20. 529 86‹
21. 653 rem.8	22. 706 rem.10	23. 5476 rem.13	
24. 1715 rem.28	25. 954 rem.70		

Test 4

1. 912	2. 4547	3. 9409	4. 115 49(
5. 141 378	6. 1217	7. 856	8. 67 789
9. 5868	10. 1813	11. 8091	12. 4263
13. 8633	14. 130 977	15. 80 464	16. 804 00:
17. 756 585	18. 822 rem.7	19. 529 rem.1	
20. 1590 rem.11	21. 1013 rem.13	22. 954 rem.66	
23. 578 rem.51	24. 3664 rem.11	25. 5033 rem.46	

EXERCISE 6 (page 6)

£	£	£	£
1. 0·48	2. 1·81	3. 1·83	4. 2·21
5. 2·26	6. 1·57$\frac{1}{2}$	7. 1·59$\frac{1}{2}$	8. 2·07$\frac{1}{2}$
9. 17·62	10. 26·71	11. 69·84	12. 58·49
13. 118·55$\frac{1}{2}$	14. 172·66$\frac{1}{2}$	15. 161·40$\frac{1}{2}$	16. 170·57$\frac{1}{2}$
17. 640·03	18. 1115·37	19. 408·61$\frac{1}{2}$	20. 523·88
21. 692·59$\frac{1}{2}$	22. 830·75$\frac{1}{2}$	23. 768·09$\frac{1}{2}$	24. 1096·00
25. 307·72	26. 224·20	27. 357·33$\frac{1}{2}$	28. 313·84$\frac{1}{2}$
29. 1420·44	30. 1945·88$\frac{1}{2}$	31. 1246·75	32. 3174·68
33. 67·35	34. 501·65$\frac{1}{2}$	35. 2113·62$\frac{1}{2}$	36. 1585·39$\frac{1}{2}$
37. 40·54	38. 88·28	39. 560·15	40. 1304·96
41. 286·17$\frac{1}{2}$	42. 146·54$\frac{1}{2}$	43. 1180·78$\frac{1}{2}$	44. 1331·86
45. 2563·50$\frac{1}{2}$	46. 2133·40$\frac{1}{2}$		

£	£	£
47. (a) 99·46	**48.** (a) 515·71	**49.** (a) 471·37$\frac{1}{2}$
(b) 126·29	(b) 549·26$\frac{1}{2}$	(b) 473·13$\frac{1}{2}$
(c) 166·26	(c) 669·01$\frac{1}{2}$	(c) 781·33$\frac{1}{2}$
(d) 67·50	(d) 1032·74$\frac{1}{2}$	(d) 1756·55
(e) 94·61	(e) 315·45$\frac{1}{2}$	(e) 1014·96
(f) 229·90	(f) 385·79	(f) 1261·57
		(g) 1205·86$\frac{1}{2}$

Total 392·01   Total 1733·99   Total 3482·39$\frac{1}{2}$

50. (a) 2047·39
  (b) 1425·23
  (c) 1097·99$\frac{1}{2}$
  (d) 938·29
  (e) 1874·93
  (f) 1418·00
  (g) 632·05
  (h) 1583·92$\frac{1}{2}$

Total 5508·90$\frac{1}{2}$

EXERCISE 7 (page 9)

£	£	£	£
1. 2·53	2. 4·04	3. 3·57	4. 10·47
5. 5·17	6. 7·16	7. 17·49$\frac{1}{2}$	8. 9·08$\frac{1}{2}$
9. 19·11$\frac{1}{2}$	10. 30·70$\frac{1}{2}$	11. 47·16	12. 35·08
13. 28·64	14. 127·08	15. 185·71	16. 331·54
17. 181·46$\frac{1}{2}$	18. 119·70$\frac{1}{2}$	19. 542·08$\frac{1}{2}$	20. 388·06$\frac{1}{2}$
21. 241·64$\frac{1}{2}$	22. 347·71$\frac{1}{2}$	23. 53·05$\frac{1}{2}$	24. 481·70$\frac{1}{2}$
25. 486·07	26. 760·18$\frac{1}{2}$	27. 582·91$\frac{1}{2}$	28. 1165·08$\frac{1}{2}$
29. 380·67$\frac{1}{2}$	30. 5·70$\frac{1}{2}$	31. 54·72	32. 16·19
33. 492·07$\frac{1}{2}$	34. 369·12$\frac{1}{2}$	35. 567·27$\frac{1}{2}$	36. 528·71$\frac{1}{2}$
37. 388·62$\frac{1}{2}$	38. 87·08$\frac{1}{2}$	39. 32·98	40. 183·35$\frac{1}{2}$
41. 278·89$\frac{1}{2}$	42. 1098·27$\frac{1}{2}$	43. 547·05	44. 807·06$\frac{1}{2}$
45. 986·49	46. 1286·15$\frac{1}{2}$	47. 1890·52$\frac{1}{2}$	48. 1628·87$\frac{1}{2}$
49. 1098·25$\frac{1}{2}$	50. 2974·35$\frac{1}{2}$	51. 1915·09	52. 1886·87
53. 888·98$\frac{1}{2}$	54. 887·81$\frac{1}{2}$	55. 1985·09	56. 1823·65$\frac{1}{2}$
57. 4582·71$\frac{1}{2}$	58. 3681·28$\frac{1}{2}$	59. 6236·79$\frac{1}{2}$	60. 8723·29$\frac{1}{2}$

EXERCISE 8 (page 11)

	(a) £	(b) £		(a) £	(b) £
1.	3·69	4·92	**2.**	16·48	20·60
3.	14·44	21·66	**4.**	41·58	47·52
5.	54·32	61·11	**6.**	49·62	57·89
7.	59·22	78·96	**8.**	63·60	71·55
9.	71·52	62·58	**10.**	28·35	37·80
11.	31·56	23·67	**12.**	49·27½	38·32½
13.	62·32½	48·47½	**14.**	98·24	85·96
15.	89·58	134·37	**16.**	108·32½	139·27½
17.	178·78½	139·05½	**18.**	86·20	103·44
19.	94·08	109·76	**20.**	142·72	115·96
21.	134·10	111·75	**22.**	102·78	119·91
23.	129·99	148·56	**24.**	183·52½	220·23
25.	324·59	440·51½	**26.**	522·02½	631·92½
27.	870·84	1052·26½	**28.**	1406·24	1274·40½
29.	778·44	1107·78	**30.**	953·12½	1486·87½
31.	1611·90½	1873·29½	**32.**	2433·37½	2000·77½
33.	1824·42	1684·08	**34.**	2173·73	2646·28
35.	5084·63	6754·21	**36.**	4143·68½	5327·59½
37.	3305·20½	4651·77	**38.**	8063·22½	10 460·40
39.	15 828·24	18 796·03½	**40.**	14 440·44	19 339·87½
41.	10 071·95½	11 705·24½	**42.**	14 989·47½	20 730·12½
43.	8363·42½	16 009·98½	**44.**	15 420·24	24 415·38
45.	12 126·35	19 443·97½	**46.**	23 495·70	30 751·72½
47.	17 818·12½	42 946·25	**48.**	26 816·96½	47 613·79½
49.	35 515·84½	30 531·16½	**50.**	47 611·52½	43 216·61½

EXERCISE 9 (page 12)

	(a) £	(b) £		(a) £	(b) £
1.	5·24	2·62	**2.**	4·22	2·11
3.	6·20	4·96	**4.**	9·38	8·04
5.	10·29	7·35	**6.**	12·18	8·12
7.	15·19	8·68	**8.**	21·06	9·36
9.	16·72	14·63	**10.**	19·26	14·98
11.	7·41	4·94	**12.**	9·88	7·41

	(a) £	(b) £		(c) £	(b) £
13.	14·07	11·72½	**14.**	28·66½	19·11
15.	12·27½	7·36½	**16.**	16·55½	11·82½
17.	34·60½	26·91½	**18.**	3·94	3·15 rem.3p

	(a) £	(b) £
19.	3·87	3·18½ rem.3½p
20.	4·93	4·15 rem.3p
21.	4·25	3·71½ rem.9p
22.	5·78	4·92 rem.10p
23.	7·74 rem.14p	8·26 rem.2p
24.	23·96	15·97 rem.9p
25.	6·26½ rem.13p	7·10 rem.14p
26.	8·24 rem.11p	14·31½ rem.4½p
27.	8·91 rem.8p	6·45 rem.14p
28.	8·64½ rem.1½p	5·37 rem.16p
29.	17·19 rem.1p	14·81½ rem.12½p
30.	7·85½ rem.7p	9·54 rem.2p
31.	11·01 rem.2p	6·19 rem.17p
32.	15·66 rem.11p	11·66 rem.19p
33.	26·38½ rem.7p	17·59 rem.7p
34.	6·09 rem.3p	4·60 rem.9p
35.	15·60 rem.9p	11·47 rem.11p
36.	14·21½ rem.5p	16·66½ rem.7½p
37.	7·68½ rem.16½p	6·24½ rem.12p
38.	10·84 rem.9p	16·07½ rem.3½p
39.	18·81½ rem.2½p	15·78 rem.3½p
40.	13·66 rem.5½p	15·87½ rem.6p
41.	18·87 rem.14½p	31·45½ rem.1p
42.	59·51 rem.1½p	30·48 rem.4½p
43.	92·80 rem.4½p	55·32 rem.20½p
44.	45·10½ rem.10p	31·22½ rem.16p
45.	153·46½ rem.4½p	82·63½ rem.4½p
46.	175·94½ rem.5½p	78·71 rem.14p
47.	45·22 rem.1½p	27·37 rem.1½p
48.	97·78 rem.12p	49·66½ rem.18½p
49.	63·09 rem.1p	120·61 rem.12p
50.	108·51 rem.23p	233·10 rem.11p

EXERCISE 10 (page 12)

1. 37	2. 57	3. 85	4. 165
5. 25	6. 28	7. 14	8. 15
9. 16	10. 16	11. 15	12. 16
13. 26	14. 24	15. 28	16. 27
17. 28	18. 27	19. 49	20. 48
21. 49	22. 49	23. ɔ0	24. 46
25. 42	26. 44	27. 45	28. 78 rem.22p

29. 62 rem.19p   30. 79 rem.7p   31. 182 rem.2p
32. 161 rem.47p   33. 115 rem.5p   34. 103 rem.45p
35. 53 rem.56p   36. 375   37. 506 rem.34p
38. 676 rem.9p   39. 524 rem.41p   40. 652 rem.45p
41. 795 rem.53p   42. 709 rem.69p   43. 220
44. 180 rem.15p   45. 177 rem.$12\frac{1}{2}$p   46. 333 rem.$4\frac{1}{2}$p
47. 381 rem.$14\frac{1}{2}$p   48. 349 rem.34p   49. 36
50. 94   51. 102 rem.65p   52. 116 rem. 84p
53. 136 rem.56p   54. 327 rem.39p   55. 314 rem.53p
56. 140 rem.45p   57. 101 rem.£2·36   58. 313 rem.35p
59. 256 rem.58p   60. 339 rem.13p

EXERCISE 11 (page 13)

£	£	£	£
1. 0·15	2. 0·36	3. 0·45	4. 0·63
5. 0·55	6. 0·70	7. 1·20	8. 1·43
9. 1·12	10. 2·86	11. 7·75	12. 1·92
13. 2·00	14. 2·43	15. 6·72	16. 4·23
17. 7·20	18. 8·82	19. 8·64	20. 10·29
21. 11·50	22. 12·25	23. 11·76	24. 17·50
25. 17·15	26. 13·72	27. 17·64	28. 12·50
29. 18·62	30. 36·75	31. 45·60	32. 86·80
33. 87·25	34. 80·82	35. 120·69	36. 156·80
37. 130·80	38. 203·28	39. 181·00	40. 131·$62\frac{1}{2}$
41. 79·74	42. 176·40	43. 156·45	44. 219·$37\frac{1}{2}$
45. 151·$87\frac{1}{2}$	46. 862·00	47. 872·$62\frac{1}{2}$	48. 1020·72
49. 968·75	50. 2032·$87\frac{1}{2}$		

EXERCISE 12 (page 14)

	£		£		£		£
1.	0·20	2.	0·18	3.	3·12	4.	0·60
	0·30		0·20		7·50		0·60
	0·26		0·66		9·20		3·50
Total	0·76	Total	1·04	Total	19·82	Total	4·70

	£		£		£		£
5.	3·36	6.	4·60	7.	2·90	8.	3·60
	1·68		3·82½		2·32½		8·60
	0·25		1·75		4·37½		1·08
Total	5·29	Total	10·17½	Total	9·60	Total	13·28

	£		£		£		£
9.	0·62½	10.	1·86	11.	2·70	12.	11·25
	1·50		1·68		3·42		21·25
	6·37½		1·12½		1·95		32·00
Total	8·50		0·82		2·92½		9·37½
		Total	5·48½	Total	10·99½	Total	73·87½

	£		£		£		£
13.	0·60	14.	3·68	15.	38·40	16.	14·20
	1·50		4·68		21·30		12·06
	1·20		2·52		7·50		51·00
	0·63		10·08		3·16		9·36
Total	3·93	Total	20·96	Total	70·36	Total	86·62

	£		£		£		£
17.	43·38	18.	15·60	19.	12·96	20.	15·00
	142·50		65·60		23·45		2·50
	48·50		42·60		35·00		1·44
	38·70		18·75		40·50		3·00
	48·96				57·75		6·00
		Total	142·55				
Total	322·04			Total	169·66	Total	27·94

9

	£		£		£		£
21.	1·50	**22.**	1·86	**23.**	17·64	**24.**	30·00
	5·25		4·70		15·54		3·91
	8·00		8·68		20·70		6·82½
	3·75		2·70		31·86		2·03
	5·60	Total	17·94		43·68		2·55
Total	24·10			Total	129·42	Total	45·31½

	£		£		£
25.	27·20	**26.**	26·45	**27.**	0·95
	37·80		3·99		8·61
	28·50		7·35		2·43
	240·00		1·74		7·44
	128·00		2·55		1·02
Total	461·50	Total	42·08		13·90½
					18·00
					6·45
				Total	58·80½

EXERCISE 13 (page 18)

Test 1

£	£	£
1. 32·11	2. 146·68	3. 47·88½
4. 72·95½	5. 303·49½	6. 61·97
7. 982·56	8. 4190·16	9. 7125·72
10. 12·41½ rem.2½p	11. 9·54 rem.3p	12. 10·99½ rem.12p
13. 188·60½ rem.6½p	14. 4·20	15. 14·87½
16. 32·00	17. 246·51	18. 189·87½

19. (a) 210·14
   (b) 335·13½
   (c) 658·77
   (d) 439·10
   (e) 622·12
   (f) 400·57½
   (g) 362·84½
   (h) 257·60½
  Total 1643·14½

20. 2·32
    2·66
   17·98
    6·65
   13·68
Total 43·29

Test 2

£	£	£
1. 192·27	2. 204·17½	3. 69·07
4. 87·16½	5. 280·55½	6. 197·22
7. 670·32	8. 1460·76	9. 3753·12
10. 6160·27½	11. 6·98 rem.3p	12. 8·90½ rem.3½p
13. 12·94½ rem.6p	14. 80·97½ rem.13½p	15. 20·50
16. 118·44	17. 268·25	

<table>
<tr><td>18.</td><td>£
21·35</td><td>19. 36</td><td>20. 316 rem.42p</td></tr>
</table>

£
18.   21·35
28·50
7·35
17·55
9·00
5·03½
5·51
Total 94·29½

Test 3

£	£	£
1. 355·20	2. 221·21	3. 647·80½
4. 267·27½	5. 554·52½	6. 850·95
7. 2168·32	8. 3172·16	9. 4375·44
10. 8·67½ rem.6½p	11. 32·14½ rem.4½p	12. 22·04 rem.10½p
13. 21·12½	14. 134·10	15. 410·62½
16. 96	17. 729 rem.20p	18. 63 rem.15p
19. 242·20	20.	

20.
	£
(a)	226·49
(b)	1008·97½
(c)	636·51
(d)	443·15½
(e)	523·65
(f)	640·36
(g)	542·60½
(h)	608·51½
Total	2315·13

Test 4

£	£	£
1. 263·51	2. 227·63	3. 1062·07
4. 162·08	5. 207·12½	6. 588·88½
7. 1247·85	8. 1773·90	9. 4152·99½

11

	£		£		£
10.	$7\cdot02\frac{1}{2}$ rem.$6\frac{1}{2}$p	**11.**	$33\cdot83$ rem.8p	**12.**	$23\cdot49$ rem.17p
13.	$35\cdot10$	**14.**	$64\cdot56$	**15.**	$317\cdot52\frac{1}{2}$
16.	180	**17.**	147 rem.39p	**18.**	99 rem.$5\frac{1}{2}$p
19.	81	**20.**			

£
```
            16·20
             1·08
             5·98½
             5·67
             2·18½
             3·15
             2·75
Total       37·02
```

EXERCISE 14 (page 21)

1. 15.6	**2.** 49.9	**3.** 105.5	**4.** 236.8
5. 1549.8	**6.** 9.94	**7.** 16.81	**8.** 31.75
9. 165.99	**10.** 1685.64	**11.** 167.905	**12.** 170.197
13. 1522.86	**14.** 194.551	**15.** 329.15	**16.** 265.455
17. 15.314	**18.** 99.486	**19.** 242.786	**20.** 157.099
21. 443.16	**22.** 148.66	**23.** 234.57	**24.** 91.636
25. 214.973	**26.** 380.1	**27.** 744.502	**28.** 213.025
29. 307.004	**30.** 209.39	**31.** 184.065	**32.** 328.169
33. 443.93	**34.** 235.8589	**35.** 289.89	**36.** 278.16
37. 1659.117	**38.** 42.801	**39.** 239.815	**40.** 292.315
41. 1127.31	**42.** 357.63	**43.** 351.975	**44.** 553.135
45. 359.97	**46.** 124.95	**47.** 286.455	**48.** 440.425
49. 690.66	**50.** 239.256	**51.** 242.724	**52.** 741.91
53. 232.971	**54.** 2089.8979	**55.** 35.858	**56.** 206.321
57. 615.5165	**58.** 109.375	**59.** 154.984	**60.** 354.69
61. 65.69612	**62.** 94.8849	**63.** 96.7275	**64.** 410.091
65. 93.358	**66.** 132.06	**67.** 143.2329	**68.** 31.685
69. 47.62	**70.** 152.091		

EXERCISE 15 (page 23)

1. 5.3	**2.** 6.5	**3.** 4.0	**4.** 25.7
5. 62.4	**6.** 80.9	**7.** 48.8	**8.** 7.4
9. 72.8	**10.** 37.7	**11.** 75.1	**12.** 67.1

13. 75.18	**14.** 11.51	**15.** 17.87	**16.** 21.09
17. 22.32	**18.** 22.78	**19.** 41.09	**20.** 71.94
21. 78.93	**22.** 60.07	**23.** 25·998	**24.** 40.987
25. 45.746	**26.** 73.938	**27.** 51.08	**28.** 4.899
29. 10.125	**30.** 23.102	**31.** 61.97	**32.** 103.93
33. 137.92	**34.** 165.83	**35.** 379.125	**36.** 652.625
37. 57.09	**38.** 116.968	**39.** 115.88	**40.** 112.94
41. 399.1877	**42.** 199.8745	**43.** 186.7243	**44.** 164.525
45. 78.012	**46.** 41.813	**47.** 80.119	**48.** 28.179
49. 41.957	**50.** 0.999	**51.** 4.12996	**52.** 10·082 41
53. 4.237	**54.** 2.8912	**55.** 7.84	**56.** 24.665
57. 11.299	**58.** 13.9	**59.** 16.935	**60.** 386.687
61. 227.974	**62.** 26.075	**63.** 6.2302	**64.** 1.21
65. 4.17	**66.** 12.25	**67.** 3.05	**68.** 85.8
69. 2.04	**70.** 6.316	**71.** 33.12	**72.** 30.53
73. 139.73	**74.** 14.9	**75.** 137.435	**76.** 21.665
77. 33.0	**78.** 15.3	**79.** 19.85	**80.** 34.9
81. 28.4	**82.** 105.6	**83.** 19.55	**84.** 23.702
85. 38.7437	**86.** 91.33	**87.** 133.6	**88.** 108.055
89. 122.66	**90.** 9.695		

EXERCISE 16 (page 24)

	(a)	(b)		(a)	(b)
1.	12.4	124.0	**2.**	276.0	27.6
3.	211.0	2110.0	**4.**	712.0	7120.0
5.	213.0	2130.0	**6.**	79.0	790.0
7.	40.0	400.0	**8.**	0.4	4.0
9.	0.06	0.6	**10.**	401.0	4010.0
11.	800.2	8002.0	**12.**	412.1	1236.3
13.	127.0	508.0	**14.**	2.7	18.9
15.	45.0	135.0	**16.**	21.3	42.6
17.	710.0	2130.0	**18.**	240.0	2160.0
19.	3710.0	11 130.0	**20.**	4910.0	34 370.0
21.	0.42	0.042	**22.**	0.079	0.79
23.	22.1	2.21	**24.**	41.7	4.17
25.	2.131	0.2131	**26.**	27.55	2.755
27.	0.029	0.0029	**28.**	0.41	0.0041
29.	0.09	0.009	**30.**	0.015	0.0015
31.	0.027	0.0027	**32.**	0.031	0.0031

	(a)	(b)		(a)	(b)
33.	0.0071	0.000 71	34.	0.004	0.0004
35.	0.0049	0.000 049	36.	0.0007	0.000 07
37.	0.000 31	0.000 031	38.	0.000 79	0.000 079
39.	0.000 02	0.000 002	40.	0.000 004 7	0.000 000 47

EXERCISE 17 (page 25)

	(a)	(b)		(a)	(b)
1.	12.6	16.8	2.	9.0	13.5
3.	20.7	27.6	4.	44.5	71.2
5.	47.0	84.6	6.	109.8	164.7
7.	128.4	171.2	8.	14.98	17.12
9.	28.32	24.78	10.	27.64	55.28
11.	72.09	152.19	12.	126.28	108.24
13.	75.36	42.39	14.	71.28	53.46
15.	683.4	321.6	16.	233.8	434.2
17.	1641.6	820.8	18.	25.34	38.01
19.	95.2	78.88	20.	105.71	71.61
21.	1092.0	1248.0	22.	1463.4	1544.7
23.	0.0957	1.479	24.	0.0266	0.462
25.	0.008	0.0096	26.	0.0484	0.0847
27.	0.2925	0.26	28.	0.3661	0.4707
29.	0.7101	0.3156	30.	15.12	17.01
31.	20.85	37.53	32.	65.52	40.95
33.	45.65	54.78	34.	21.75	20.3
35.	30.24	24.57	36.	49.86	58.17
37.	196.56	237.51	38.	0.828	1.5272
39.	1031.69	981.97	40.	636.35	616.77
41.	421.59	708.63	42.	373.5	468.12
43.	166.43	91.63	44.	9.5625	10.0125
45.	3.5921	2.5356	46.	3.743	9.259
47.	75.99	73.308	48.	73.549	84.721
49.	41.064	32.568	50.	83.142	67.944
51.	28.971	44.955	52.	2.6999	133.574
53.	3.2778	58.272	54.	5.6289	9.1227
55.	56.451	34.599	56.	93.786	170.346
57.	15.9012	16.8477	58.	194.999	9.8595
59.	7.96	24.1984	60.	16.0524	153.972
61.	1.906 59	3.38517	62.	0.1914	0.216 92
63.	0.205 92	0.244 53	64.	0.198 15	0.224 57
65.	0.383 76	0.3198	66.	0.592 99	0.499 36

	(a)	(b)		(a)	(b)
67.	0.792 49	0.709 07	68.	547.75	635.39
69.	2025.17	1446.55	70.	0.113 03	0.062 23

EXERCISE 18 (page 26)

1. 0.5625	2. 7.4925	3. 6.942	4. 19.6425
5. 0.0215	6. 0.131 25	7. 0.000 84	8. 0.001 41
9. 1.8792	10. 0.006 615	11. 0.018 26	12. 0.000 42
13. 0.16	14. 0.002 25	15. 0.04	16. 0.006 48
17. 0.011 424	18. 0.000 56	19. 0.016 05	20. 0.001 68

EXERCISE 19 (page 26)

	(a)	(b)		(a)	(b)
1.	81.6	8.16	2.	9.43	0.943
3.	8.97	0.897	4.	4.98	0.498
5.	9.39	0.939	6.	8.76	0.876
7.	9.399	0.9399	8.	0.976	0.0976
9.	0.423	0.0423	10.	0.372	0.0372
11.	8.5	5.1	12.	0.085	0.051
13.	1.2096	0.5184	14.	2.49	4.15
15.	16.9	8.45	16.	15.7	7.85
17.	0.62	0.31	18.	0.44	0.22
19.	1.04	0.52	20.	1.34	0.67
21.	6.48	3.24	22.	3.92	1.96
23.	23.25	11.625	24.	2.68	1.34
25.	0.0648	0.0324	26.	0.224	2.24
27.	7.8	15.6	28.	7.7	53.9
29.	1000.0	1400.0	30.	13.0	1.3
31.	1.6	0.16	32.	2.7	0.27
33.	833.0	83.3	34.	154.0	15.4
35.	3.6	36.0	36.	48.29	4.829
37.	3338.0	333.8	38.	2649.0	264.9
39.	23.28	232.8	40.	345.3	34.53
41.	36.74	367.4	42.	387.5	38.75
43.	29.84	2.984	44.	13.79	1.379
45.	182.9	1829.0	46.	17.34	1.734
47.	1652.0	165.2	48.	156.4	15.64
49.	14.76	147.6	50.	1.385	13.85

	(a)	(b)		(a)	(b)
51.	3.8	6.333	52.	4.5	3.857
53.	5.444	44.545	54.	7.25	0.621
55.	0.62	5.166	56.	2.531	25.315
57.	1.861	1.448	58.	1.545	1.496
59.	3.311	1.986	60.	3.174	3.644
61.	1.2	0.747	62.	1.209	1.049
63.	1.216	3.648	64.	0.694	0.808
65.	1.533	0.953	66.	2.059	0.597
67.	1.075	41.476	68.	23.16	19.965
69.	0.152	1.164	70.	1.187	47.0
71.	12.333	77.894	72.	25.211	34.423
73.	3.352	38.256	74.	85.860	29.574
75.	11.793	10.482	76.	1.446	1.414
77.	4.211	2.915	78.	6.156	5.420
79.	10.333	11.986	80.	70.328	62.826
81.	102.776	88.242	82.	1071.904	968.172
83.	12.552	1.100	84.	0.057	0.051
85.	0.184	0.237	86.	1.001	1.563
87.	50.817	143.689	88.	2.420	4.733
89.	6.291	4141.666	90.	84.84	1631.538

EXERCISE 20 (page 28)

1. 0.025	2. 1.666	3. 0.5	4. 2.5
5. 0.001 25	6. 0.25	7. 0.02	8. 0.128
9. 0.025	10. 20.833	11. 6.730	12. 0.06
13. 6.864	14. 0.242	15. 126.0	16. 18.461
17. 43.076	18. 0.002	19. 54.0	20. 0.380
21. 57.142	22. 53.666	23. 6.363	24. 1170.0
25. 9493.75			

EXERCISE 21 (page 29)

Test 1

1. 22.78	2. 70.318	3. 171.044	4. 2.125
5. 7.755	6. 79.913	7. 0.777	8. 3.649
9. 2.645	10. 0.399 84	11. 98.952	12. 9.543
13. 17.141	14. 7.318	15. 54.833	16. 5.986

17. 90.141	18. 55.22	19. 19.979	20. 77.801
21. 36.745	22. 9.09	23. 50.617	24. 12.733
25. 49.5			

Test 2

1. 22.74	2. 88.979	3. 184.802	4. 631.197
5. 2.105	6. 18.155	7. 35.9811	8. 8.19
9. 4.116	10. 10.73	11. 4.5496	12. 0.391 95
13. 33.508	14. 4780.588	15. 10.319	16. 17.175
17. 395.652	18. 5.349	19. 47.423	20. 14.13
21. 61.169	22. 19.14	23. 117.22	24. 42.8916
25. 74.8216			

Test 3

1. 442.02	2. 112.095	3. 931.691	4. 248.007
5. 22.4	6. 59.925	7. 231.124	8. 2.523
9. 19.11	10. 5.859	11. 3.9342	12. 0.031 584
13. 0.002 556	14. 14.395	15. 5.9	16. 0.382
17. 2.171	18. 4.6	19. 10.413	20. 104.835
21. 39.275	22. 71.365	23. 31.17	24. 2.1515
25. 0.0202			

Test 4

1. 116.665	2. 292.498	3. 799.368	4. 1202.264
5. 42.98	6. 57.63	7. 552.088	8. 664.944
9. 25.92	10. 7.476	11. 0.5929	12. 0.187
13. 1.81875	14. 38.666	15. 3.037	16. 6.422
17. 17.692	18. 93.06	19. 64.255	20. 221.06

21. (a)   684.956
   (b)   463.377
   (c)   501·207
   (d)   713·329
   (e)   910.426
   (f)   175.688
   (g)   360.075
   (h)   916.68
Total   2362.869

EXERCISE 22 (page 31)

1. 15.5 cm	2. 31.1 cm	3. 46.3 cm
4. 52.1 cm	5. 1.01 m	6. 3.91 m
7. 5.75 mᵢ	8. 7.31 m	9. 10.65 m
10. 11.63 m	11. 23.33 m	12. 1.775 km
13. 2.425 km	14. 1.605 km	15. 2.69 km
16. 1.28 kg	17. 1.899 kg	18. 2.714 kg
19. 1.974 kg	20. 1.338 tonnes	21. 1.993 tonnes
22. 2.943 tonnes	23. 1.271 litres	24. 2.232 litres
25. 2.533 litres	26. 5.101 km	27. 38.996 m
28. 49.6 kg	29. 14.467 km	30. 21.98 km
31. 34.361 km	32. 271.983 km	33. 11.264 kg
34. 59.004 kg	35. 234.802 kg	36. 249.791 tonnes
37. 499.66 tonnes	38. 32.189 g	39. 340.03 g
40. 32.1 m	41. 1107.225 m	42. 684.844 m
43. 37.196 m	44. 64.217 m	45. 248.902 m
46. 1.84575 km	47. 684.239 kg	48. 1.470 15 tonnes
49. 2.475 litres	50. 1.691 litres	51. 2.617 litres
52. 2.249 litres	53. 15 700 cm³	54. 51 911 cm³
55. 109 178 cm³	56. 4.304 m	57. 10.388 m
58. 8.835 m	59. 5.59 m	60. 6.95 kg

EXERCISE 23 (page 34)

1. 22.3 cm	2. 54.7 cm	3. 42.7 cm
4. 890 mm	5. 1620 mm	6. 2912 mm
7. 7125 mm	8. 7645 mm	9. 30.9 km
10. 13 640 m	11. 20 480 m	12. 135 050 m
13. 2.142 kg	14. 1.913 kg	15. 2250 g
16. 4800 g	17. 4745 g	18. 6600 kg
19. 4165 kg	20. 7245 kg	21. 0.5377 tonnes
22. 0.681 25 tonnes	23. 0.2694 km	24. 1.074 25 km
25. 0.735 295 km	26. 1.911 m	27. 11.673 m
28. 13.926 km	29. 3.78 g	30. 11.67 litres
31. 6.59 kg	32. 8.769 km	33. 6.208 g
34. 6.948 km	35. 20.147 km	36. 57.157 km
37. 13.75 kg	38. 84.955 kg	39. 98.115 kg
40. 7.645 tonnes	41. 28.525 tonnes	42. 5.974 m
43. 17.975 m	44. 53.775 m	45. 274.325 m
46. 582.565 km	47. 185.753 km	48. 127.475 kg
49. 199.825 kg	50. 299.775 kg	

EXERCISE 24 (page 35)

	(a)	(b)		(a)	(b)
1.	232 mm	261 mm	2.	861 mm	1107 mm
3.	105.6 cm	66 cm	4.	151.9 cm	130.2 cm
5.	3.924 m	3.052 m	6.	4.944 m	3.708 m
7.	7.407 m	5.761 m	8.	1908 g	1484 g
9.	2.512 kg	1.884 kg	10.	2.961 kg	3.807 kg
11.	5.184 kg	3.456 kg	12.	238 m	272 m
13.	667 m	812 m	14.	2.688 km	3.584 km
15.	5.355 km	7.245 km	16.	6.762 km	9.996 km
17.	13.631 km	17.752 km	18.	13.197 km	8.964 km
19.	26.775 km	19.975 km	20.	25.382 km	35.224 km
21.	6.814 tonnes	10.011 tonnes			
22.	17.484 tonnes	14.136 tonnes			
23.	14.007 tonnes	22.701 tonnes			
24.	1215 mm	1080 mm	25.	3312 mm	4140 mm
26.	4592 mm	6232 mm	27.	11 155 mm	15 520 mm
28.	7475 mm	10 166 mm	29.	11.375 m	13.65 m
30.	11.116 m	13.895 m	31.	905.28 m	1216.47 m
32.	1048.77 m	902.43 m	33.	1926 cm	2568 cm
34.	25 606 cm	31 093 cm	35.	76 132 cm	92 446 cm
36.	2.2005 km	1.5485 km	37.	2.0288 km	2.853 km
38.	1.3104 tonnes	1.5288 tonnes			
39.	1.8834 tonnes	2.4528 tonnes			
40.	1.9573 tonnes	2.3805 tonnes			
41.	2.7765 tonnes	3.5169 tonnes			
42.	1.8202 kg	2.2513 kg	43.	3.8963 kg	5.3056 kg
44.	4.2262 kg	5.3449 kg	45.	5.817 kg	7.756 kg
46.	1.656 litres	2.52 litres	47.	3.906 litres	5.022 litres
48.	7.371 litres	6.669 litres	49.	10.272 litres	12.626 litres
50.	25.125 litres	31.5 litres			

EXERCISE 25 (page 36)

1. 9 mm	2. 16 mm	3. 29 mm	4. 38 mm
5. 29 mm	6. 33 mm	7. 47 mm	8. 39 mm
9. 43 mm	10. 34 mm	11. 21 cm	12. 45 cm
13. 39 cm	14. 25 cm	15. 38 cm	16. 310 mm
17. 310 mm	18. 310 mm	19. 210 m	20. 450 m

19

21. 280 m	22. 340 m	23. 33 g	24. 38 g
25. 49 g	26. 49 g	27. 39 g	28. 31 g
29. 0.048 kg	30. 0.49 kg	31. 0.38 kg	32. 0.39 kg

33. 0.043 tonnes   34. 0.047 tonnes   35. 0.043 tonnes
36. 0.047 tonnes   37. 0.047 tonnes   38. 0.033 km

39. 0.042 km	40. 0.045 km	41. 0.49 km	42. 0.48 km
43. 0.147 km	44. 49 dm	45. 43 dm	46. 47 dm
47. 49 dm	48. 48 dm	49. 0.49 m	50. 0.49 m

51. 0.27 m   52. 0.481 litres   53. 0.572 litres
54. 0.645 litres   55. 0.814 litres   56. 0.678 litres

57. 280 cm	58. 490 cm	59. 460 cm	60. 490 cm
61. 430 kg	62. 1470 kg	63. 470 kg	64. 1840 kg

65. 2300 kg

EXERCISE 26 (page 37)

Test 1

1. 96 cm   2. 2.408 m   3. 5.27 m   4. 1.788 tonnes
5. 1.943 kg   6. 2.547 km   7. 22.114 km   8. 20.111 km
9. 4500 g   10. 2480 g   11. 0.785 75 tonnes
12. 0.587 95 km   13. 1.692 m   14. 9.765 kg   15. 9.328 tonnes
16. 75.6 m   17. 0.464 88 km   18. 4.074 litres
19. 7 litres   20. 490 mm   21. 0.049 km   22. 0.044 tonnes
23. 490 m   24. 460 litres   25. 471.5 kg

Test 2

1. 223.6 cm   2. 2.327 m   3. 5.9 m   4. 3.746 m
5. 1.833 tonnes   6. 2.165 kg   7. 2.026 litres   8. 29 910 m
9. 26.32 g   10. 73.176 km   11. 27.798 tonnes
12. 4200 g   13. 3800 m   14. 0.229 75 km
15. 1.316 m   16. 0.342 44 km   17. 8.892 kg   18. 3.304 litres
19. 70.47 km   20. 1.484 g   21. 38 270 cm   22. 48 cm
23. 0.47 litres   24. 2600 m   25. 4.8 litres

Test 3

1. 64.275 km   2. 166.2 cm   3. 1.932 litres
4. 2.236 m   5. 11.462 kg   6. 2.87 tonnes
7. 2.788 litres   8. 64.76 g   9. 346.331 km

10. 48.303 tonnes 11. 4200 g 12. 20 060 m
13. 86 850 m 14. 126 cm 15. 3.572 m 16. 6.132 kg
17. 0.0749 18. 30 240 cm 19. 2.9563 g 20. 29 520 m
21. 50 960 cm 22. 4.1 cm 23. 0.39 m 24. 1700 kg
25. 1.2 litres

EXERCISE 27 (page 39)

	(a)	(b)	(c)		(a)	(b)	(c)
1.	$\frac{3}{2}$	$\frac{8}{3}$	$\frac{17}{4}$	2.	$\frac{5}{2}$	$\frac{14}{3}$	$\frac{10}{3}$
3.	$\frac{13}{4}$	$\frac{11}{3}$	$\frac{11}{4}$	4.	$\frac{19}{4}$	$\frac{15}{2}$	$\frac{8}{3}$
5.	$\frac{28}{5}$	$\frac{15}{4}$	$\frac{19}{2}$	6.	$\frac{34}{5}$	$\frac{9}{2}$	$\frac{26}{3}$
7.	$\frac{43}{6}$	$\frac{31}{11}$	$\frac{29}{4}$	8.	$\frac{17}{6}$	$\frac{37}{4}$	$\frac{37}{5}$
9.	$\frac{89}{10}$	$\frac{47}{5}$	$\frac{31}{6}$	10.	$\frac{85}{11}$	$\frac{59}{7}$	$\frac{33}{5}$
11.	$\frac{107}{12}$	$\frac{71}{6}$	$\frac{81}{16}$	12.	$\frac{86}{11}$	$\frac{107}{16}$	$\frac{39}{5}$
13.	$\frac{173}{18}$	$\frac{68}{5}$	$\frac{35}{6}$	14.	$\frac{227}{12}$	$\frac{125}{8}$	$\frac{53}{6}$
15.	$\frac{149}{12}$	$\frac{94}{9}$	$\frac{43}{7}$	16.	$\frac{160}{13}$	$\frac{87}{8}$	$\frac{73}{7}$
17.	$\frac{80}{9}$	$\frac{125}{8}$	$\frac{81}{8}$	18.	$\frac{115}{9}$	$\frac{53}{5}$	$\frac{80}{7}$
19.	$\frac{71}{10}$	$\frac{112}{9}$	$\frac{107}{8}$	20.	$\frac{43}{5}$	$\frac{150}{11}$	$\frac{109}{9}$
21.	$\frac{85}{9}$	$\frac{147}{10}$	$\frac{157}{8}$	22.	$\frac{161}{9}$	$\frac{173}{12}$	$\frac{175}{9}$
23.	$\frac{203}{12}$	$\frac{177}{10}$	$\frac{143}{8}$	24.	$\frac{239}{15}$	$\frac{256}{13}$	$\frac{161}{10}$
25.	$\frac{119}{8}$	$\frac{187}{12}$	$\frac{119}{8}$	26.	$\frac{201}{10}$	$\frac{391}{20}$	$\frac{103}{10}$
27.	$\frac{499}{16}$	$\frac{202}{15}$	$\frac{115}{12}$	28.	$\frac{677}{16}$	$\frac{347}{20}$	$\frac{101}{12}$
29.	$\frac{131}{3}$	$\frac{253}{15}$	$\frac{97}{10}$	30.	$\frac{182}{3}$	$\frac{309}{20}$	$\frac{113}{15}$
31.	$\frac{83}{10}$	$\frac{253}{16}$	$\frac{99}{10}$	32.	$\frac{39}{5}$	$\frac{181}{21}$	$\frac{101}{20}$
33.	$\frac{59}{6}$	$\frac{255}{16}$	$\frac{29}{3}$	34.	$\frac{277}{15}$	$\frac{167}{20}$	$\frac{123}{20}$
35.	$\frac{130}{7}$	$\frac{214}{23}$	$\frac{187}{20}$	36.	$1\frac{1}{2}$	$1\frac{1}{2}$	$1\frac{3}{4}$
37.	$2\frac{2}{7}$	$1\frac{1}{3}$	$6\frac{1}{2}$	38.	$2\frac{1}{2}$	$4\frac{3}{4}$	$2\frac{1}{8}$
39.	$5\frac{2}{3}$	$2\frac{1}{4}$	$4\frac{1}{2}$	40.	$4\frac{3}{4}$	$1\frac{4}{5}$	$1\frac{5}{6}$
41.	$8\frac{1}{2}$	$2\frac{1}{5}$	$3\frac{2}{5}$	42.	$1\frac{7}{12}$	$1\frac{11}{12}$	$1\frac{8}{19}$
43.	$2\frac{5}{7}$	$4\frac{1}{7}$	$2\frac{4}{5}$	44.	$6\frac{5}{6}$	$5\frac{5}{6}$	$5\frac{4}{7}$
45.	$6\frac{1}{3}$	$9\frac{1}{2}$	$15\frac{2}{3}$	46.	$7\frac{7}{8}$	$7\frac{1}{8}$	$14\frac{1}{6}$
47.	$6\frac{1}{8}$	$9\frac{2}{9}$	$7\frac{5}{12}$	48.	$4\frac{1}{12}$	$15\frac{1}{2}$	$10\frac{6}{7}$
49.	$17\frac{4}{5}$	$31\frac{2}{3}$	$18\frac{1}{3}$	50.	$4\frac{4}{5}$	$13\frac{2}{3}$	$9\frac{7}{8}$
51.	$4\frac{9}{10}$	$12\frac{1}{4}$	$20\frac{4}{5}$	52.	$14\frac{5}{6}$	$21\frac{1}{6}$	$29\frac{2}{5}$
53.	$19\frac{1}{5}$	$10\frac{7}{10}$	$8\frac{9}{10}$	54.	$16\frac{4}{5}$	$4\frac{9}{25}$	$4\frac{9}{24}$
55.	$12\frac{7}{10}$	$14\frac{9}{10}$	$2\frac{7}{20}$	56.	$3\frac{8}{25}$	$11\frac{2}{9}$	$13\frac{8}{11}$
57.	$28\frac{5}{7}$	$19\frac{6}{13}$	$16\frac{2}{3}$	58.	$18\frac{8}{19}$	$6\frac{1}{16}$	$6\frac{1}{14}$
59.	$6\frac{7}{12}$	$6\frac{1}{3}$	$2\frac{1}{5}$	60.	$6\frac{13}{14}$	$6\frac{3}{16}$	$13\frac{4}{11}$
61.	$17\frac{4}{11}$	$36\frac{4}{11}$	$35\frac{2}{3}$	62.	$5\frac{4}{25}$	$15\frac{11}{12}$	$18\frac{2}{9}$
63.	$8\frac{17}{25}$	$66\frac{2}{3}$	$8\frac{1}{11}$	64.	$27\frac{3}{11}$	$18\frac{9}{10}$	$8\frac{13}{20}$
65.	$12\frac{14}{25}$	$33\frac{1}{3}$	$9\frac{1}{2}$				

EXERCISE 28 (page 40)

1. $\frac{1}{2}$	2. $\frac{2}{3}$	3. $\frac{2}{3}$	4. $\frac{3}{4}$	5. $\frac{8}{9}$
6. $\frac{3}{4}$	7. $\frac{3}{4}$	8. $\frac{8}{9}$	9. $\frac{5}{11}$	10. $\frac{13}{17}$
11. $\frac{4}{5}$	12. $\frac{9}{13}$	13. $\frac{5}{6}$	14. $\frac{4}{7}$	15. $\frac{7}{10}$
16. $\frac{13}{17}$	17. $\frac{9}{16}$	18. $\frac{3}{40}$	19. $\frac{7}{15}$	20. $\frac{4}{5}$
21. $\frac{3}{5}$	22. $\frac{6}{7}$	23. $\frac{3}{4}$	24. $\frac{7}{10}$	25. $\frac{3}{4}$
26. $\frac{5}{6}$	27. $\frac{4}{5}$	28. $\frac{2}{9}$	29. $\frac{3}{8}$	30. $\frac{2}{3}$
31. $\frac{2}{3}$	32. $\frac{2}{5}$	33. $\frac{3}{4}$	34. $\frac{9}{10}$	35. $\frac{4}{7}$
36. $\frac{2}{5}$	37. $\frac{15}{16}$	38. $\frac{3}{4}$	39. $\frac{2}{3}$	40. $\frac{17}{20}$
41. $\frac{5}{11}$	42. $\frac{9}{70}$	43. $\frac{9}{104}$	44. $\frac{11}{13}$	45. $\frac{187}{249}$
46. $1\frac{3}{5}$	47. $1\frac{5}{7}$	48. $2\frac{1}{2}$	49. $1\frac{1}{2}$	50. $1\frac{1}{5}$
51. $1\frac{2}{3}$	52. $2\frac{2}{5}$	53. $1\frac{2}{3}$	54. $2\frac{2}{3}$	55. $1\frac{5}{6}$
56. $1\frac{1}{3}$	57. $1\frac{1}{3}$	58. $3\frac{4}{5}$	59. $1\frac{9}{17}$	60. $1\frac{3}{5}$
61. $2\frac{2}{5}$	62. $2\frac{1}{3}$	63. $4\frac{1}{5}$	64. $2\frac{3}{5}$	65. $8\frac{2}{5}$
66. $1\frac{1}{8}$	67. $2\frac{4}{5}$	68. $1\frac{1}{2}$	69. $1\frac{2}{3}$	70. $2\frac{1}{6}$
71. $3\frac{3}{5}$	72. $1\frac{1}{4}$	73. $6\frac{2}{3}$	74. $2\frac{2}{3}$	75. $2\frac{1}{2}$
76. $12\frac{2}{5}$	77. $2\frac{1}{7}$	78. $1\frac{1}{7}$	79. $4\frac{1}{2}$	80. 3
81. $1\frac{5}{7}$	82. $5\frac{1}{2}$	83. $9\frac{3}{4}$	84. $2\frac{2}{3}$	85. $8\frac{3}{5}$
86. 12	87. $1\frac{17}{20}$	88. $2\frac{47}{50}$	89. $8\frac{4}{7}$	90. $26\frac{2}{3}$

EXERCISE 29 (page 41)

£	£	£	£
1. 2·42	2. 1·88	3. 1·69	4. 2·28
5. 6·92	6. 9·63	7. 5·76	8. 5·80
9. 13·65	10. 4·59	11. 32·77$\frac{1}{2}$	12. 17·61
13. 48·16	14. 114·48	15. 12·66	16. 33·67
17. 47·56$\frac{1}{2}$	18. 64·84$\frac{1}{2}$	19. 90·67$\frac{1}{2}$	20. 56·50
21. 83·70	22. 90·79	23. 149·63	24. 235·41
25. 102·42	26. 500 m	27. 750 m	28. 1400 m
29. 750 m	30. 300 g	31. 800 g	32. 2800 g
33. 3150 g	34. 2730 g	35. 6012.5 g	36. 5775 mm
37. 10 1750 mm	38. 2100 kg	39. 5525 kg	40. 3420 cm^3
41. 4472 cm^3	42. 6842 cm^3		43. 6987 cm^3
44. 4655 g	45. 8640 g	46. 15 130 kg	47. 12 920 m
48. 4230 m	49. 3080 kg	50. 3360 cm^3	
51. 3960 m	52. 14 210 m	53. 15 840 g	54. 17 020 g
55. 1806 m	56. 0.0783 km	57. 0.1081 km	58. £190·13$\frac{1}{2}$
59. £368·12$\frac{1}{2}$	60. £403·32$\frac{1}{2}$		

22

EXERCISE 30 (page 42)

1. $\frac{5}{6}$ 2. $\frac{3}{4}$ 3. $\frac{7}{10}$ 4. $\frac{2}{3}$ 5. $\frac{9}{14}$
6. $\frac{5}{8}$ 7. $\frac{3}{5}$ 8. $\frac{11}{18}$ 9. $\frac{7}{12}$ 10. $\frac{17}{30}$
11. $\frac{11}{20}$ 12. $\frac{27}{50}$ 13. $\frac{7}{12}$ 14. $\frac{9}{20}$ 15. $\frac{5}{12}$
16. $\frac{3}{8}$ 17. $\frac{7}{20}$ 18. $\frac{1}{3}$ 19. $\frac{3}{10}$ 20. $\frac{29}{100}$
21. $\frac{11}{15}$ 22. $\frac{14}{15}$ 23. $\frac{1}{2}$ 24. $\frac{16}{21}$ 25. $\frac{17}{24}$
26. $\frac{19}{28}$ 27. $\frac{24}{35}$ 28. $1\frac{1}{12}$ 29. $\frac{19}{20}$ 30. $\frac{7}{12}$
31. $\frac{7}{8}$ 32. $\frac{3}{4}$ 33. $\frac{9}{10}$ 34. $\frac{5}{6}$ 35. $\frac{3}{4}$
36. $1\frac{1}{4}$ 37. $1\frac{1}{4}$ 38. $1\frac{3}{8}$ 39. $1\frac{1}{2}$ 40. $1\frac{3}{10}$
41. $1\frac{1}{3}$ 42. $1\frac{13}{20}$ 43. $1\frac{1}{4}$ 44. $1\frac{11}{24}$ 45. $1\frac{1}{2}$
46. $1\frac{11}{20}$ 47. $1\frac{13}{16}$ 48. $1\frac{9}{16}$ 49. $1\frac{19}{36}$ 50. $1\frac{5}{12}$
51. $2\frac{1}{6}$ 52. $2\frac{9}{20}$ 53. $2\frac{1}{4}$ 54. $1\frac{17}{30}$ 55. $1\frac{23}{24}$
56. $2\frac{1}{4}$ 57. $2\frac{7}{60}$ 58. $2\frac{37}{60}$ 59. $2\frac{23}{36}$ 60. $2\frac{17}{24}$
61. $2\frac{15}{44}$ 62. $3\frac{3}{4}$ 63. $5\frac{8}{15}$ 64. $10\frac{9}{14}$ 65. $5\frac{5}{6}$
66. $5\frac{9}{10}$ 67. $8\frac{3}{5}$ 68. $3\frac{5}{6}$ 69. $8\frac{1}{4}$ 70. $7\frac{5}{8}$
71. $11\frac{1}{2}$ 72. $6\frac{17}{20}$ 73. $6\frac{7}{16}$ 74. $8\frac{11}{20}$ 75. $8\frac{1}{4}$
76. $11\frac{7}{18}$ 77. $8\frac{7}{24}$ 78. $13\frac{8}{15}$ 79. $8\frac{23}{36}$ 80. $7\frac{7}{22}$
81. $8\frac{1}{2}$ 82. $31\frac{1}{4}$ 83. $6\frac{5}{8}$ 84. 11 85. $5\frac{11}{16}$
86. $10\frac{6}{7}$ 87. $27\frac{14}{25}$ 88. $12\frac{2}{5}$ 89. $12\frac{3}{8}$ 90. $10\frac{11}{24}$
91. $11\frac{11}{24}$ 92. $19\frac{11}{18}$ 93. $20\frac{19}{30}$ 94. $16\frac{11}{12}$ 95. $16\frac{7}{24}$
96. $11\frac{7}{24}$ 97. $10\frac{13}{36}$ 98. $7\frac{11}{20}$ 99. $9\frac{47}{50}$ 100. $7\frac{17}{60}$

EXERCISE 31 (page 43)

1. $\frac{1}{4}$ 2. $\frac{1}{2}$ 3. $\frac{3}{10}$ 4. $\frac{1}{6}$ 5. $\frac{5}{14}$
6. $\frac{3}{8}$ 7. $\frac{7}{18}$ 8. $\frac{2}{5}$ 9. $\frac{5}{12}$ 10. $\frac{1}{60}$
11. $\frac{9}{20}$ 12. $\frac{1}{4}$ 13. $\frac{1}{2}$ 14. $\frac{1}{2}$ 15. $\frac{5}{18}$
16. $\frac{3}{22}$ 17. $\frac{3}{8}$ 18. $\frac{1}{15}$ 19. $\frac{1}{12}$ 20. $\frac{1}{9}$
21. $\frac{1}{6}$ 22. $\frac{1}{60}$ 23. $\frac{8}{35}$ 24. $\frac{25}{36}$ 25. $\frac{19}{48}$
26. $\frac{19}{60}$ 27. $\frac{23}{60}$ 28. $\frac{1}{140}$ 29. $\frac{25}{126}$ 30. $1\frac{1}{6}$
31. $2\frac{5}{12}$ 32. $2\frac{3}{8}$ 33. $2\frac{3}{8}$ 34. $6\frac{7}{16}$ 35. $4\frac{1}{2}$
36. $2\frac{9}{20}$ 37. $2\frac{5}{12}$ 38. $3\frac{1}{4}$ 39. $2\frac{5}{18}$ 40. $3\frac{1}{2}$
41. $1\frac{5}{8}$ 42. $3\frac{7}{24}$ 43. $2\frac{1}{2}$ 44. $2\frac{9}{20}$ 45. $3\frac{37}{72}$
46. $\frac{3}{4}$ 47. $1\frac{3}{4}$ 48. $\frac{11}{12}$ 49. $1\frac{9}{10}$ 50. $2\frac{1}{30}$
51. $1\frac{5}{16}$ 52. $\frac{9}{20}$ 53. $1\frac{11}{16}$ 54. $1\frac{13}{16}$ 55. $5\frac{8}{15}$
56. $1\frac{2}{3}$ 57. $1\frac{7}{10}$ 58. $\frac{27}{35}$ 59. $\frac{19}{45}$ 60. $\frac{5}{8}$
61. $2\frac{7}{9}$ 62. $3\frac{27}{44}$ 63. $4\frac{29}{50}$ 64. $7\frac{1}{2}$ 65. $1\frac{1}{3}$
66. $1\frac{1}{6}$ 67. $6\frac{9}{20}$ 68. $6\frac{3}{10}$ 69. $6\frac{61}{100}$ 70. $2\frac{1}{2}$
71. $2\frac{3}{4}$ 72. $3\frac{1}{4}$ 73. $6\frac{3}{8}$ 74. $3\frac{4}{5}$ 75. $2\frac{5}{8}$
76. $6\frac{7}{10}$ 77. $1\frac{19}{24}$ 78. $4\frac{5}{6}$ 79. $5\frac{9}{28}$ 80. $8\frac{1}{3}$

81. $12\frac{1}{30}$ 82. $2\frac{5}{8}$ 83. $1\frac{3}{4}$ 84. $5\frac{1}{2}$ 85. $10\frac{3}{4}$
86. $1\frac{7}{8}$ 87. $22\frac{11}{30}$ 88. $5\frac{7}{9}$ 89. $5\frac{1}{3}$ 90. $15\frac{1}{4}$
91. $17\frac{5}{8}$ 92. $5\frac{1}{10}$ 93. $9\frac{7}{12}$ 94. $2\frac{1}{4}$ 95. 7
96. $3\frac{11}{20}$ 97. $7\frac{21}{50}$ 98. $7\frac{11}{16}$ 99. $5\frac{11}{16}$ 100. $7\frac{3}{4}$

EXERCISE 32 (page 44)

1. $1\frac{1}{2}$ 2. 4 3. 6 4. 12 5. 14
6. 24 7. 30 8. 18 9. 4 10. 9
11. $1\frac{1}{3}$ 12. $5\frac{1}{4}$ 13. $5\frac{1}{3}$ 14. $4\frac{4}{7}$ 15. 6
16. $4\frac{4}{5}$ 17. $\frac{1}{2}$ 18. $\frac{1}{3}$ 19. $\frac{7}{45}$ 20. $\frac{1}{8}$
21. $\frac{1}{4}$ 22. $\frac{1}{3}$ 23. $\frac{7}{36}$ 24. $\frac{1}{9}$ 25. $\frac{1}{12}$
26. $\frac{1}{2}$ 27. $\frac{2}{5}$ 28. $\frac{1}{4}$ 29. $\frac{7}{18}$ 30. $\frac{4}{7}$
31. $\frac{2}{3}$ 32. $\frac{3}{5}$ 33. $\frac{3}{4}$ 34. $\frac{1}{2}$ 35. $\frac{2}{3}$
36. $\frac{3}{4}$ 37. $\frac{2}{5}$ 38. $\frac{7}{18}$ 39. $\frac{1}{2}$ 40. $\frac{7}{20}$
41. $\frac{4}{7}$ 42. $\frac{2}{3}$ 43. $1\frac{7}{8}$ 44. $5\frac{5}{8}$ 45. $16\frac{1}{5}$
46. 11 47. $3\frac{9}{16}$ 48. 9 49. 20 50. $5\frac{1}{7}$
51. $2\frac{1}{4}$ 52. $35\frac{1}{3}$ 53. $10\frac{5}{16}$ 54. 4 55. 6
56. $12\frac{3}{8}$ 57. $2\frac{1}{5}$ 58. $7\frac{7}{15}$ 59. 9 60. $22\frac{1}{2}$
61. $6\frac{3}{16}$ 62. $3\frac{7}{16}$ 63. $10\frac{2}{3}$ 64. $23\frac{2}{5}$ 65. $4\frac{2}{3}$
66. $10\frac{2}{5}$ 67. $4\frac{5}{7}$ 68. $6\frac{2}{3}$ 69. 6 70. $19\frac{5}{7}$
71. $9\frac{3}{4}$ 72. $6\frac{1}{6}$ 73. $8\frac{5}{8}$ 74. 12 75. 12
76. $8\frac{7}{30}$ 77. $17\frac{11}{12}$ 78. $29\frac{6}{7}$ 79. $20\frac{2}{9}$ 80. $32\frac{1}{2}$
81. $44\frac{1}{3}$ 82. $12\frac{4}{7}$ 83. 24 84. 48 85. 36
86. $3\frac{1}{3}$ 87. $5\frac{1}{4}$ 88. $10\frac{1}{2}$ 89. 5 90. $22\frac{3}{16}$
91. $2\frac{2}{15}$ 92. $1\frac{1}{2}$ 93. $14\frac{1}{4}$ 94. 17 95. 3
96. $1\frac{1}{2}$ 97. 24 98. $5\frac{5}{8}$ 99. $6\frac{3}{10}$ 100. $16\frac{1}{2}$
101. $4\frac{1}{8}$ 102. $1\frac{1}{3}$ 103. $3\frac{3}{4}$ 104. $2\frac{5}{8}$ 105. 14
106. 14 107. $16\frac{1}{4}$ 108. $14\frac{7}{16}$ 109. $11\frac{3}{8}$ 110. $37\frac{5}{7}$
111. $18\frac{1}{5}$ 112. $30\frac{5}{8}$ 113. $36\frac{3}{4}$ 114. $33\frac{19}{27}$ 115. $100\frac{15}{16}$
116. 4 117. 7 118. $18\frac{1}{3}$ 119. $60\frac{3}{8}$ 120. $42\frac{3}{16}$

EXERCISE 33 (page 45)

1. $\frac{2}{3}$ 2. $\frac{20}{27}$ 3. $1\frac{1}{5}$ 4. $\frac{8}{15}$ 5. $\frac{15}{28}$
6. $\frac{16}{21}$ 7. $\frac{3}{4}$ 8. $\frac{7}{8}$ 9. $\frac{2}{3}$ 10. $\frac{3}{4}$
11. $\frac{3}{4}$ 12. $1\frac{5}{27}$ 13. $1\frac{1}{5}$ 14. $\frac{27}{32}$ 15. $1\frac{1}{14}$
16. $1\frac{1}{8}$ 17. $3\frac{1}{4}$ 18. $2\frac{5}{8}$ 19. $1\frac{2}{9}$ 20. $\frac{9}{10}$
21. $1\frac{5}{8}$ 22. $1\frac{2}{5}$ 23. $\frac{1}{9}$ 24. $\frac{1}{18}$ 25. $\frac{1}{8}$
26. $\frac{3}{17}$ 27. $\frac{1}{30}$ 28. $\frac{1}{58}$ 29. $\frac{2}{45}$ 30. $\frac{1}{21}$

31. $\frac{1}{33}$ 32. $\frac{4}{35}$ 33. $\frac{3}{49}$ 34. $1\frac{2}{5}$ 35. $2\frac{1}{6}$

36. $2\frac{1}{2}$ 37. $3\frac{1}{5}$ 38. 2 39. $\frac{1}{2}$ 40. 1

41. 3 42. 4 43. $2\frac{1}{2}$ 44. $1\frac{1}{5}$ 45. $8\frac{7}{8}$

46. $1\frac{31}{33}$ 47. $1\frac{1}{5}$ 48. 2 49. $1\frac{1}{4}$ 50. $1\frac{1}{2}$

51. $\frac{3}{4}$ 52. $3\frac{2}{3}$ 53. $1\frac{7}{9}$ 54. $2\frac{2}{11}$ 55. $3\frac{5}{6}$

56. $2\frac{1}{4}$ 57. $1\frac{1}{5}$ 58. $1\frac{2}{7}$ 59. $3\frac{7}{10}$ 60. $1\frac{1}{20}$

61. $3\frac{5}{6}$ 62. $2\frac{6}{9}$ 63. $2\frac{2}{7}$ 64. $1\frac{7}{12}$ 65. $3\frac{1}{14}$

66. 30 67. $4\frac{3}{4}$ 68. 2 69. $2\frac{2}{3}$ 70. $3\frac{5}{7}$

71. 18 72. $3\frac{5}{7}$ 73. $\frac{4}{7}$ 74. $4\frac{3}{4}$ 75. $\frac{5}{8}$

76. $2\frac{4}{9}$ 77. $\frac{2}{5}$ 78. $3\frac{3}{5}$ 79. $2\frac{1}{7}$ 80. $2\frac{2}{7}$

81. 75 82. $\frac{9}{17}$ 83. $\frac{8}{15}$ 84. 36 85. 6

86. 16 87. 2 88. $\frac{1}{2}$ 89. $\frac{5}{7}$ 90. $1\frac{26}{99}$

91. $1\frac{2}{5}$ 92. 6 93. $\frac{28}{33}$ 94. $2\frac{1}{4}$ 95. $\frac{22}{35}$

96. 3 97. $\frac{25}{44}$ 98. 6 99. $5\frac{2}{11}$ 100. $\frac{2}{5}$

101. $2\frac{5}{6}$ 102. 1 103. 5 104. $\frac{14}{15}$ 105. $\frac{99}{112}$

106. 2 107. $\frac{3}{7}$ 108. $1\frac{17}{18}$ 109. $4\frac{2}{7}$ 110. $5\frac{31}{64}$

111. $2\frac{25}{28}$ 112. $2\frac{23}{27}$ 113. 3 114. $3\frac{3}{5}$ 115. $15\frac{13}{15}$

116. $\frac{9}{10}$ 117. $1\frac{13}{32}$ 118. $\frac{3}{4}$ 119. 72 120. $\frac{1}{12}$

121. $\frac{1}{8}$ 122. $\frac{1}{2}$ 123. 4 124. $1\frac{3}{4}$ 125. $\frac{4}{9}$

126. $\frac{15}{32}$ 127. $10\frac{4}{5}$ 128. $1\frac{3}{4}$ 129. $\frac{8}{35}$ 130. $\frac{3}{4}$

131. $\frac{189}{1024}$ 132. $1\frac{4}{5}$ 133. $7\frac{1}{2}$ 134. $1\frac{3}{5}$ 135. $\frac{2}{9}$

136. $\frac{10}{63}$ 137. $\frac{45}{112}$ 138. $\frac{31}{56}$ 139. $27\frac{3}{7}$ 140. $1\frac{1}{3}$

EXERCISE 34 (page 46)

	£		£		£		£
1.	9·00	2.	9·00	3.	27·00	4.	36·00
5.	109·00	6.	72·00	7.	8·00	8.	13·00
9.	21·00	10.	12·00	11.	32·00	12.	54·00
13.	48·00	14.	39·00	15.	84·00	16.	3·50
17.	8·50	18.	11·50	19.	3·75	20.	6·75
21.	45·00	22.	63·00	23.	135·00	24.	28·00
25.	54·00	26.	77·00	27.	3·00	28.	8·00
29.	14·00	30.	18·00	31.	40·00	32.	10·00
33.	15·00	34.	24·00	35.	96·00	36.	55·00
37.	85·00	38.	168·00	39.	189·00	40.	14·00
41.	154·00	42.	57·75	43.	87·75	44.	4·00
45.	48·00	46.	60·00	47.	126·00	48.	96·00
49.	91·00	50.	128·00	51.	91·00	52.	147·00
53.	81·00	54.	132·00	55.	273·00	56.	168·00
57.	540·00	58.	121·50	59.	529·00	60.	293·75

£	£	£	£
61. 123·25	**62.** 178·25	**63.** 222·75	**64.** 82·00
65. 352·50	**66.** 306·25	**67.** 166·50	**68.** 181·50
69. 342·50	**70.** 626·50		

EXERCISE 35 (page 47)

1. 0.5	**2.** 0.25	**3.** 0.75	**4.** 0.125
5. 0.375	**6.** 0.625	**7.** 0.875	**8.** 0.2
9. 0.6	**10.** 0.8	**11.** 0.4	**12.** 0.3
13. 0.1	**14.** 0.7	**15.** 0.9	**16.** 0.05
17. 0.15	**18.** 0.35	**19.** 0.45	**20.** 0.55
21. 0.65	**22.** 0.85	**23.** 0.95	**24.** 0.0625
25. 0.1875	**26.** 0.3125	**27.** 0.4375	**28.** 0.5625
29. 0.6875	**30.** 0.8125	**31.** 0.9375	**32.** 0.031 25
33. 0.093 75	**34.** 0.156 25	**35.** 0.218 75	**36.** 0.281 25
37. 0.343 75	**38.** 0.406 25	**39.** 0.468 75	**40.** 0.531 25
41. 0.593 75	**42.** 0.656 25	**43.** 0.718 75	**44.** 0.843 75
45. 0.781 25	**46.** 0.968 75	**47.** 0.01	**48.** 0.03
49. 0.04	**50.** 0.12	**51.** 0.17	**52.** 0.08
53. 0.21	**54.** 0.31	**55.** 0.19	**56.** 0.28
57. 0.44	**58.** 0.47	**59.** 0.36	**60.** 0.81
61. 0.52	**62.** 0.68	**63.** 0.91	**64.** 0.007
65. 0.047	**66.** 0.009	**67.** 0.43	**68.** 0.793
69. 0.39	**70.** 1.5	**71.** 4.75	**72.** 5.875
73. 9.4375	**74.** 10.4	**75.** 10.6	**76.** 8.4
77. 4.1	**78.** 3.7	**79.** 2.2	**80.** 4.9
81. 7.04	**82.** 8.44	**83.** 10.075	**84.** 8.3125
85. 5.5625	**86.** 4.175	**87.** 3.343 75	**88.** 2.225
89. 1.9	**90.** 4.8	**91.** 5.4	**92.** 7.03
93. 8.17	**94.** 4.16	**95.** 0.642	**96.** 0.785
97. 0.285	**98.** 0.666	**99.** 0.333	**100.** 0.111
101. 0.444	**102.** 0.555	**103.** 0.777	**104.** 0.523
105. 0.166	**106.** 0.833	**107.** 0.857	**108.** 0.888
109. 0.272	**110.** 0.090	**111.** 0.909	**112.** 0.076
113. 0.538	**114.** 0.692	**115.** 0.846	**116.** 0.473
·117. 0.214	**118.** 0.357	**119.** 0.928	**120.** 0.933

EXERCISE 36 (page 49)

1. $\frac{1}{2}$	**2.** $\frac{1}{4}$	**3.** $\frac{3}{4}$	**4.** $\frac{1}{10}$	**5.** $\frac{3}{10}$
6. $\frac{13}{20}$	**7.** $\frac{19}{20}$	**8.** $\frac{7}{20}$	**9.** $\frac{9}{20}$	**10.** $\frac{3}{20}$

11. $\frac{17}{20}$ 12. $\frac{9}{10}$ 13. $\frac{191}{200}$ 14. $\frac{1}{5}$ 15. $\frac{37}{100}$
16. $\frac{2}{5}$ 17. $\frac{3}{5}$ 18. $\frac{7}{10}$ 19. $\frac{4}{5}$ 20. $\frac{1}{20}$
21. $\frac{1}{100}$ 22. $\frac{3}{100}$ 23. $\frac{7}{25}$ 24. $\frac{7}{100}$ 25. $\frac{9}{50}$
26. $\frac{9}{100}$ 27. $\frac{19}{100}$ 28. $\frac{1}{25}$ 29. $\frac{27}{100}$ 30. $\frac{3}{50}$
31. $\frac{9}{25}$ 32. $\frac{2}{25}$ 33. $\frac{11}{25}$ 34. $\frac{16}{25}$ 35. $\frac{11}{20}$
36. $\frac{93}{100}$ 37. $\frac{47}{100}$ 38. $\frac{5}{8}$ 39. $\frac{1}{8}$ 40. $\frac{3}{8}$
41. $\frac{29}{40}$ 42. $\frac{17}{40}$ 43. $\frac{33}{40}$ 44. $\frac{37}{40}$ 45. $\frac{23}{40}$
46. $\frac{27}{40}$ 47. $\frac{22}{25}$ 48. $\frac{51}{400}$ 49. $\frac{149}{400}$ 50. $\frac{913}{2000}$
51. $\frac{189}{400}$ 52. $\frac{197}{400}$ 53. $\frac{313}{400}$ 54. $\frac{301}{400}$ 55. $\frac{147}{200}$
56. $\frac{179}{200}$ 57. $\frac{7}{8}$ 58. $\frac{87}{200}$ 59. $\frac{119}{200}$ 60. $\frac{167}{200}$
61. $\frac{147}{200}$ 62. $\frac{171}{200}$ 63. $\frac{151}{200}$ 64. $\frac{187}{200}$ 65. $\frac{421}{500}$
66. $\frac{181}{200}$ 67. $\frac{163}{200}$ 68. $\frac{143}{200}$ 69. $\frac{13}{40}$ 70. $\frac{69}{200}$
71. $\frac{79}{200}$ 72. $\frac{39}{200}$ 73. $\frac{37}{200}$ 74. $\frac{33}{200}$ 75. $\frac{5}{16}$
76. $3\frac{17}{80}$ 77. $2\frac{33}{80}$ 78. $1\frac{49}{80}$ 79. $1\frac{13}{16}$ 80. $4\frac{73}{80}$
81. $7\frac{61}{80}$ 82. $8\frac{1687}{2000}$ 83. $8\frac{1887}{2000}$ 84. $9\frac{379}{400}$ 85. $10\frac{139}{400}$
86. $9\frac{99}{400}$ 87. $3\frac{359}{400}$ 88. $4\frac{49}{80}$ 89. $7\frac{1}{32}$ 90. $5\frac{29}{2000}$

EXERCISE 37 (page 49)

Test 1

1. $6\frac{1}{4}$ 2. $9\frac{3}{8}$ 3. $7\frac{13}{30}$ 4. $1\frac{1}{6}$ 5. $1\frac{1}{6}$
6. $1\frac{25}{72}$ 7. $\frac{9}{14}$ 8. $3\frac{3}{4}$ 9. 5 10. $1\frac{1}{4}$
11. $1\frac{1}{5}$ 12. $2\frac{8}{15}$ 13. $2\frac{13}{16}$ 14. $5\frac{7}{10}$ 15. £16·0
16. £6·37$\frac{1}{2}$ 17. £468 18. £102 19. £7·28 20. $4\frac{1}{2}$
21. $2\frac{5}{21}$ 22. £5·78 23. $5\frac{2}{3}$ 24. $6\frac{11}{20}$ 25. 4340 cm

Test 2

1. $7\frac{7}{8}$ 2. $11\frac{13}{24}$ 3. $10\frac{17}{24}$ 4. $2\frac{1}{6}$ 5. $1\frac{3}{4}$
6. $3\frac{7}{18}$ 7. $\frac{1}{2}$ 8. $14\frac{3}{4}$ 9. $4\frac{1}{8}$ 10. $2\frac{7}{8}$
11. $\frac{1}{2}$ 12. $3\frac{4}{5}$ 13. $1\frac{3}{4}$ 14. $2\frac{2}{3}$ 15. $3\frac{1}{2}$
16. 3 17. £52 18. £150 19. 3750 m 20. $\frac{19}{39}$
21. $1\frac{7}{10}$ 22. 2 tonnes 23. $2\frac{9}{16}$ 24. $2\frac{7}{12}$ 25. $2\frac{47}{48}$

Test 3

1. $3\frac{3}{4}$ 2. $8\frac{3}{4}$ 3. $2\frac{1}{2}$ 4. $\frac{15}{16}$ 5. $2\frac{23}{60}$
6. $5\frac{7}{30}$ 7. $6\frac{19}{60}$ 8. $5\frac{1}{4}$ 9. 11 10. $3\frac{4}{27}$
11. $\frac{2}{3}$ 12. 2 13. $1\frac{11}{14}$ 14. 4 15. £39
16. £144 17. £13·09$\frac{1}{2}$ 18. $\frac{19}{35}$ 19. $\frac{3}{4}$ 20. £14·77
21. 1260 cm 22. 394 100 m 23. $5\frac{5}{6}$
24. $2\frac{3}{4}$ 25. $5\frac{3}{8}$

Test 4

1. $6\frac{13}{24}$ 2. $10\frac{61}{90}$ 3. $13\frac{41}{60}$ 4. $1\frac{1}{6}$ 5. $\frac{13}{24}$
6. $1\frac{9}{40}$ 7. 12 8. $5\frac{1}{4}$ 9. $6\frac{3}{32}$ 10. $1\frac{2}{3}$
11. $1\frac{53}{72}$ 12. 5 13. $5\frac{4}{5}$ 14. $\frac{8}{21}$ 15. $3\frac{1}{2}$
16. £9·50 17. £22·19 18. £63 19. £104·50 20. £114·75
21. 45 000 cm 22. $5\frac{179}{360}$ 23. $3\frac{3}{16}$ 24. $4\frac{9}{10}$ 25. 8

Test 5

1. $3\frac{15}{16}$ 2. $7\frac{41}{80}$ 3. $8\frac{25}{48}$ 4. $2\frac{1}{3}$ 5. $3\frac{23}{60}$
6. $3\frac{13}{60}$ 7. 5 8. $12\frac{3}{16}$ 9. 2 10. $2\frac{2}{9}$
11. $1\frac{3}{5}$ 12. $2\frac{8}{11}$ 13. $8\frac{2}{5}$ 14. $1\frac{2}{3}$ 15. £74·75
16. £85·25 17. £14·50 18. $\frac{3}{7}$. 19. $\frac{37}{55}$ 20. £41·26$\frac{1}{2}$
21. £56·81 22. 1200 kg 23. $5\frac{23}{24}$ 24. $5\frac{37}{54}$
25. $7\frac{29}{42}$

Test 6

1. $2\frac{11}{16}$ 2. $1\frac{7}{10}$ 3. $6\frac{11}{15}$ 4. $8\frac{7}{36}$ 5. $4\frac{139}{140}$
6. $9\frac{13}{120}$ 7. $1\frac{37}{60}$ 8. $3\frac{1}{6}$ 9. 2 10. $5\frac{1}{4}$
11. $\frac{11}{12}$ 12. $9\frac{3}{5}$ 13. £74 14. £145 15. £19·20
16. £287·50 17. $\frac{3}{8}$ 18. $\frac{1}{36}$ 19. £15·01$\frac{1}{2}$ 20. 975 cm
21. 150 800 m 22. 3 23. $\frac{27}{64}$ 24. $5\frac{1}{3}$
25. $4\frac{39}{80}$

EXERCISE 38 (page 52)

	Perimeter	*Area*		*Perimeter*	*Area*
1.	40 mm	96 mm²	2.	40 mm	84 mm²
3.	64 mm	252 mm²	4.	74 m	336 m²
5.	70 m	304 m²	6.	6.6 m	2.52 m²
7.	8.8 m	4.48 m²	8.	15.2 m	14.28 m²
9.	18.8 m	21.28 m²	10.	24 km	32.76 km²
11.	22.4 km	30.72 km²	12.	11.1 m	7.245 m²
13.	13.3 m	10.2 m²	14.	17.36 m	18.13 m²
15.	17.5 m	15.625 m²			

16. 67 cm²
17. (a) 2030 m²   (f) 116 m²
    (b) 1450 m²   (g) 156 m²
    (c) 812 m²   (h) 4292 m²
    (d) 120 m²   (i) 388 m²
    (e) 116 m²   (j) 4800 m²

18. Area = 152 cm², perimeter = 120 cm
19. (*a*) 360 m²   (*e*) 720 m²
    (*b*) 360 m²   (*f*) 910 m²
    (*c*) 60 m²   (*g*) 122 m
    (*d*)130 m²
20. (*a*) 0.6125 m²   (*c*) 0.575 m²
    (*b*) 0.6125 m²   (*d*) 1.8 m²

EXERCISE 39 (page 55)

1. 101 cm²	2. 258 cm²	3. 440 m²
4. 9 cm²	5. 10 cm²	6. 384 cm²
7. 528 cm²	8. 1320 m²	9. 14 cm²
10. 1072 m²	11. 476 m²	12. 348 m²

EXERCISE 40 (page 57)

1. 100 mm²	2. 400 mm²	3. 756 mm²
4. 874 mm²	5. 1404 mm²	6. 1845 mm²
7. 1692 mm²	8. 6.4 m²	9. 7.44 m²
10. 6.24 m²	11. 4.18 m²	12. 2.205 m²
13. 2.915 m²	14. 2.7 m²	15. 5.25 m²
16. 6.8 m²	17. 3.33 m²	18. 5.415 m²
19. 11.088 m²	20. 9.975 m²	21. 9.79 m²
22. 2.4975 m²	23. 2.875 m²	24. 2.475 m²
25. 5.265 m²	26. 24 mm	27. 24 mm
28. 30 m	29. 48 m	30. 40 mm
31. 38 mm	32. 84 mm	33. 88 mm
34. 96 mm	35. 74 mm	36. 406 mm²
37. 855 cm²	38. 300.5 mm²	39. 5.215 m²
40. 6570.5 mm²		

EXERCISE 41 (page 59)

£	£	£	£
1. 1·44	2. 3·51	3. 5·04	4. 6·40
5. 10·80	6. 7·92	7. 23·04	8. 23·52
9. 24·75	10. 52·92	11. 30·24	12. 47·84
13. 66·96	14. 126·28	15. 90·72	16. 129·03

£	£	£	£
17. 10·92	**18.** 15·12	**19.** 50·57½	**20.** 109·06
21. 82·21½	**22.** 108·00	**23.** 126·87½	**24.** 294·98
25. 80·00	**26.** 126·00	**27.** 45·00	**28.** 216·00
29. 3·84	**30.** 8·64	**31.** 14·00	**32.** 23·52
33. 23·04	**34.** 600·60	**35.** 922·56	**36.** 284·16
37. 49·50			

£	£	£
38. (a) 52·92	**39.** (a) 45·36	**40.** (a) 1·84
(b) 35·60	(b) 45·36	(b) 1·76
(c) 136·50	(c) 17·85	(c) 3·78
(d) 225·02	(d) 1215·20	(d) 88·62
	(e) 1323·77	

EXERCISE 42 (page 62)

1. 1.71 m²	**2.** 3.21 m²	**3.** 3.84 m²	**4.** 3.48 m²
5. 1.355 m²	**6.** 1.48 m²	**7.** 0.82 m²	**8.** 2.02 m²
9. 1.005 m²	**10.** 0.91 m²	**11.** 1.66 m²	**12.** 2.49 m²
13. 2.03 m²	**14.** 2.075 m²	**15.** 2.82 m²	**16.** 1.895 m²
17. 1.39 m²	**18.** 1.515 m²	**19.** 1.93 m²	**20.** 2.655 m²
21. 53.2 m²	**22.** 38.32 m²	**23.** 70.28 m²	**24.** 90.56 m²
25. 74.52 m²	**26.** 118.19 m²	**27.** 115 m²	**28.** 128.88 m²
29. 90.24 m²	**30.** 122.76 m²	**31.** 114.33 m²	**32.** 91.42 m²
33. 97.908 m²	**34.** 69.31 m²	**35.** 65.9 m²	

£	£	£	£
36. 17·92	**37.** 17·28	**38.** 21·34	**39.** 17·89
40. 19·89	**41.** 25·57	**42.** 22·70	**43.** 29·44
44. 33·83	**45.** 38·91	**46.** 39·57	**47.** 38·27
48. 42·82	**49.** 36·61	**50.** 49·37	

EXERCISE 43 (page 63)

1. (a) 500 m² **2.** (a) 2200 mm² **3.** (a) 6300 mm²
                    (b) £0·88   (c) £1·26

4. (b) 1680 m² **5.** (b) 53 200 mm² **6.** (b) 80 m²
   (c) £210·00   (c) 53 200 mm²   (c) 703 m²

7. (a) 55.98 m²
   (b) 7.02 m²
   (c) £45·36
   (d) 17.25 m²
   (e) £44·78 to the nearest penny
   (f) £26·28
   (g) £116·42

8. (a) 46.2 m²
   (b) £64·68

9. (a) 81.6 m²
   (b) 35.7 m²
   (c) 101.1 m²
   (d) 218.4 m²
   (e) £20·22
   (f) £7·14
   (g) £97·92
   (h) £125·28

EXERCISE 44 (page 67)

1. 38.4 m²	2. 36.8 m²	3. 43.2 m²	4. 41.4 m²
5. 44.16 m²	6. 49.44 m²	7. 47.5 m²	8. 47.5 m²
9. 40.8 m²	10. 50.5 m²	11. 38.64 m²	12. 48.48 m²
13. 43.68 m²	14. 40.56 m²	15. 48.25 m²	16. 40.02 m²
17. 39.56 m²	18. 44.62 m²	19. 34.31 m²	20. 36.19 m²
21. 39.2 m²	22. 39.36 m²	23. 39.528 m²	24. 40.992 m²
25. 40.504 m²	26. 47.628 m²	27. 48.884 m²	28. 45.08 m²
29. 43.788 m²	30. 52.768 m²	31. 39.672 m²	32. 42.968 m²
33. 45.528 m²	34. 39.936 m²	35. 44.82 m²	36. 51.015 m²
37. 48.95 m²	38. 49.29 m²	39. 50.88 m²	40. 65.72 m²

EXERCISE 45 (page 68)

Test 1

	Perimeter	Area		Perimeter	Area
1.	21 m	27 m²	2.	17 m	17.5 m²
3.	29 m	52 m²	4.	26 m	42.24 m²
5.	34 m	72.25 m²	6.	22.9 m	32.5 m²
7.	60.8 m	220.8 m²	8.	63.2 m	246.4 m²
9.	55.8 m	173.9 m²	10.	99 m	607.5 m²
11.	500 m²	12. 74 m²	13.	525 mm²	14. 360 m²

15. £2·34
16. (a) £57·46
    (b) £54·61
    (c) £67·32
    (d) £111·38
    (e) £104·72
    (f) £81·14
    (g) £8·60
    (h) £5·75
    (i) 37 m²
    (j) 36.5 m²
    (k) 40 m²
    (l) 46.5 m²
    (m) £14·50

	Perimeter	*Area*		*Perimeter*	*Area*
1.	22 m	30 m²	2.	14.5 m	13.11 m²
3.	18.1 m	20.335 m²	4.	19.7 m	24.225 m²
5.	17.1 m	18.27 m²	6.	16.25 m	16.5 m²
7.	17.9 m	19.92 m²	8.	16.3 m	16.5 m²
9.	19.3 m	23.1 m²	10.	24.1 m	35.7 m²

11. 234 m²   12. 1000 mm²   13. 17 mm²   14. 1056 m²

15. 110 m²   16. 499 m²   17. 140 m²

18. (*a*) 42 m²   (*e*) 133 m²
   (*b*) 28.5 m²   (*f*) £24·57
   (*c*) 37 m²   (*g*) £6·32 to the nearest penny
   (*d*) 47 m   (*h*) £5·71 to the nearest penny

EXERCISE 46 (page 73)

1. 2^5	2. 3^4	3. 7^4	4. $2^3 3^2$
5. $4^3 7^3$	6. $5^3 3^3$	7. $2^3 a^3$	8. b^5
9. $7q^6$	10. $2^3 g^3$	11. y^5	12. $4^3 x^3$
13. $x^3 y^2$	14. $a^4 b^2$	15. $2^2 a^3 b^2$	16. $3^3 a^2 b^3$
17. $5^3 a^3 y^3$	18. $5^2 a^3 b^3$	19. $3^4 (xy)^3$	20. $2^3 3^3 (xy)^2 ab$
21. 8	22. 27	23. 256	24. 81
25. 32	26. 1	27. 1000	28. 125
29. 32	30. 72	31. 432	32. 972
33. 512	34. 3888	35. 576	36. 125
37. 9	38. 343	39. 64	40. 36
41. 432	42. 432	43. 192	44. 72
45. 432	46. 4	47. 15	48. 96
49. 20	50. 24	51. 16	52. 8
53. 3	54. 16	55. 128	56. 75
57. 800	58. 1200	59. 400	60. 172 800

EXERCISE 47 (page 74)

1. x^7	2. x^9	3. t^{10}	4. y^7	5. d^{21}
6. x^7	7. x^{15}	8. m^{18}	9. n^6	10. x^{10}
11. $6x^7$	12. $12y^5$	13. $21x^9$	14. $48x^{11}$	15. $273x^{12}$

16. $24x^6$ 17. $6x^{24}$ 18. $12x^{18}$ 19. $24x^{14}$ 20. A^{11}
21. x^7 22. x 23. y^8 24. L^6 25. M^4
26. x^3 27. g^{13} 28. h^6 29. x^7 30. $4x^2$
31. $5x^4$ 32. $3y^7$ 33. $3M^7$ 34. $2x^8$ 35. $4t^8$
36. $2a^2b$ 37. $2a^6b^4$ 38. $2x^2y^4$ 39. $3x^2m$ 40. $4x^6$

41. $\dfrac{10x}{3}$ 42. $4a^2x^2$ 43. $2a^2b^5$ 44. $3xy$ 45. $9xy^6$

46. $6x^{10}$ 47. $8x^{15}$ 48. $9x^3$ 49. $2y^{12}$ 50. $7x^2y^7$
51. $18x^3$ 52. $6x^{12}$ 53. $2x^4$ 54. $6x^7$ 55. $28y^3$

56. $4y^2$ 57. $\dfrac{5y^4}{2}$ 58. $\dfrac{20y^2}{13}$ 59. $4y^4$ 60 $\frac{4}{5}x$

61. $2a^2$ 62. $2ax$ 63. $\frac{3}{5}a^2b^2c$ 64. $\dfrac{x^5y}{3}$ 65. $\dfrac{y}{2x}$

66. $\dfrac{xb^3}{2}$ 67. $\dfrac{S^2x}{m}$ 68. $\dfrac{3ab}{20}$ 69. $5ab^2xy^2$

70. $\dfrac{3xya}{11}$

EXERCISE 48 (page 75)

1. $+6$ 2. -15 3. $+2$ 4. -5 5. $+2$
6. $+3$ 7. -17 8. -29 9. -4 10. $+28$
11. -7 12. 0 13. $+83$ 14. $+59$ 15. -21
16. $+30$ 17. $+14$ 18. 0 19. -63 20. $+83$
21. $+215$ 22. $+28$ 23. -103 24. $+317$ 25. 0
26. -1 27. $+6$ 28. -14 29. $+39$ 30. -42
31. 0 32. $+20$ 33. -44 34. -88 35. $+270$
36. -82 37. -41 38. $+21$ 39. $-7\frac{1}{4}$ 40. $-25\frac{7}{8}$
41. $+37.3$ 42. $+8\frac{1}{12}$ 43. -2.2 44. $+0.05$ 45. -5.03
46. -2 47. $+11$ 48. $+15$ 49. $+30$ 50. -43
51. -67 52. $+101$ 53. -7 54. -12 55. -121
56. -78 57. -18 58. -92 59. -120 60. $+66$
61. -61 62. $+74$ 63. $+44$ 64. -6 65. $+382$
66. $+14$ 67. $+50$ 68. -69 69. $+54$ 70. $+103$
71. -9 72. -67 73. $+\frac{1}{4}$ 74. $-2\frac{1}{4}$ 75. $+7\frac{1}{6}$
76. $+7.6$ 77. -2.2 78. $-11\frac{1}{12}$ 79. -3.15 80. $+63.75$
81. $-8\frac{1}{6}$ 82. $+16.9$ 83. $-1\frac{1}{10}$ 84. $-\frac{3}{4}$ 85. $+7.87$
86. 0 87. -16 88. $+3\frac{7}{20}$ 89. $+1\frac{3}{4}$ 90. -1.25

33

EXERCISE 49 (page 76)

1.	-24	2.	$+56$	3.	-42	4.	-68	5.	-56
6.	$+108$	7.	-154	8.	$+153$	9.	$+105$	10.	$+105$
11.	-117	12.	$+238$	13.	-2	14.	$-\frac{1}{4}$	15.	$+5\frac{5}{8}$
16.	-12	17.	$+12$	18.	$-22\frac{1}{2}$	19.	$-7\frac{3}{4}$	20.	$-10\frac{1}{2}$
21.	-1134	22.	$+693$	23.	$+4410$	24.	$+987$	25.	$-24\frac{1}{2}$
26.	-98	27.	$+213$	28.	-112	29.	$+9$	30.	$+100$
31.	$+144$	32.	-8	33.	-16	34.	$+48$	35.	$+36$
36.	$+432$	37.	-432	38.	-1024	39.	$+2700$	40.	-36
41.	-36	42.	$+224$	43.	$+384$	44.	$+96$	45.	$+7056$
46.	$+4$	47.	-15	48.	$+8$	49.	$+30$	50.	-40
51.	-9	52.	-32	53.	$+24$	54.	-8	55.	-96
56.	-8	57.	-252	58.	$+16$	59.	$+18$	60.	$+16$
61.	$+24$	62.	$+96$	63.	-36	64.	$+36$	65.	$+16$
66.	$+36$	67.	$+144$	68.	$+64$	69.	$+36$	70.	$+25$
71.	$+24$	72.	-24	73.	-64	74.	-96	75.	-216

EXERCISE 50 (page 78)

1.	$+5$	2.	-2	3.	$+5$	4.	-2	5.	-7
6.	$+5$	7.	$+6$	8.	$+3$	9.	$+10$	10.	$+7$
11.	$+14$	12.	-3	13.	-5	14.	-2	15.	-12
16.	-2	17.	-15	18.	-16	19.	$+9$	20.	-20
21.	$+240$	22.	-41	23.	-14	24.	-5	25.	$+4\frac{4}{5}$
26.	$+5$	27.	-5	28.	-1.4	29.	-0.9	30.	$+2\frac{1}{2}$
31.	-3	32.	-8	33.	$+9$	34.	-11	35.	-11
36.	$+3\frac{1}{4}$	37.	$-5\frac{2}{3}$	38.	$+1\frac{8}{21}$	39.	$-1\frac{13}{14}$	40.	$-2\frac{10}{13}$
41.	$+3$	42.	-4	43.	-8	44.	-4	45.	-5
46.	-4	47.	$-4a$	48.	$-7x$	49.	$+3y$	50.	$-5xy$
51.	-3	52.	-6	53.	-15	54.	$+\frac{1}{2}$	55.	$+1\frac{1}{3}$
56.	$+6$	57.	$+8$	58.	$+5$	59.	$+\frac{4}{9}$	60.	$+4$
61.	$-\frac{8}{9}$	62.	$-\frac{1}{4}$	63.	$-\frac{3}{16}$	64.	$+\frac{6}{7}$	65.	$-2\frac{2}{7}$

EXERCISE 51 (page 80)

1.	10	2.	6	3.	16	4.	-1	5.	1
6.	-6	7.	-7	8.	2	9.	18	10.	11
11.	2	12.	14	13.	$3x$	14.	$5x$	15.	$3x$
16.	$9x$	17.	$-x$	18.	$-3x^2$	19.	$2x$	20.	$3x^2$

34

21. $3a^2 + 6a$ 22. $8a^2 - 16a$ 23. $-48a^2 - 12a$
24. $-8a^3 + 4a^2$ 25. $6a^2 - 6a + 12$ 26. $4x^3 + 2x^2 - 8x$
27. $18x^3 + 3x^2 + 6x$ 28. $-4x^3 + 2x^2 - 4x$ 29. $3x^3 - 4x^2 + 3x + 8$
30. $4x^4 - 6x^3 + 2x^2 - 8x$ 31. 5   32. -12
33. -4 34. 57   35. 40   36. -28 37. -72
38. -24 39. -20 40. -3 41. 2   42. 8
43. 32   44. 0   45. -36 46. -32 47. -12
48. 4   49. 48   50. 24

EXERCISE 52 (page 81)

1. 9   2. 12   3. 13   4. -10 5. -12
6. -6 7. -10 8. -9 9. -9 10. 0
11. 4   12. 21   13. 13   14. 8   15. 27
16. 35   17. -18 18. 48   19. -126 20. 176
21. $3\frac{3}{4}$ 22. $13\frac{2}{7}$ 23. $47\frac{3}{4}$ 24. 16   25. 54
26. $1\frac{2}{3}$ 27. $\frac{1}{2}$ 28. $-\frac{3}{10}$ 29. 25   30. $40\frac{5}{6}$
31. 2   32. 12   33. -5 34. 4   35. -4
36. -5 37. 2   38. -3 39. 4   40. 20
41. 5   42. 0.03   43. 2   44. 12   45. 7
46. 1   47. -2 48. 8   49. 1   50. 5
51. $7\frac{2}{3}$ 52. $-\frac{5}{7}$ 53. $4\frac{2}{5}$ 54. $-1\frac{3}{4}$ 55. $-22\frac{1}{2}$
56. $\frac{1}{4}$ 57. 0   58. $1\frac{1}{3}$ 59. -2 60. 3
61. $12\frac{1}{2}$ 62. -5 63. $\frac{3}{4}$ 64. 2   65. $\frac{3}{4}$
66. -54 67. $1\frac{3}{8}$ 68. $\frac{1}{6}$ 69. -8 70. $\frac{5}{9}$
71. $2\frac{1}{2}$ 72. $\frac{1}{2}$

EXERCISE 53 (page 82)

Test 1

1. 24 243   2. 18 140   3. 3796   4. 3181
5. 10 763   6. 105 032   7. 5083   8. 567
9. 1911   10. 6873   11. 3348   12. 42 245
13. 12 792   14. 710 rem.1   15. 686 rem.1
16. 165 rem.11   17. 492 rem.25   18. 375 rem.6
19. 52 087   20. 101 327   21. 61 131   22. 8555
23. 3609   24. 5929   25. 4544

Test 2

1. 1073	**2.** 4753	**3.** 6063	**4.** 10 175
5. 367 983	**6.** 19 614	**7.** 33 918	**8.** 16 768
9. 55 117	**10.** 552 rem.3		**11.** 303 rem.12
12. 270 rem.5	**13.** 614 rem.5		**14.** 327 rem.25
15. 816 rem.53	**16.** 89 247		**17.** 3927
18. 34 929	**19.** 18 624	**20.** 19 991	**21.** 836
22. 6679	**23.** 4984	**24.** 7190	**25.** 2295

Test 3

1. 1225	**2.** 3589	**3.** 3103	**4.** 15 651
5. 12 792	**6.** 178 352	**7.** 24 680	**8.** 68 119
9. 33 211	**10.** 29 257	**11.** 4573	**12.** 525
13. 18 689	**14.** 57 987	**15.** 236 rem.7	**16.** 292 rem.1?
17. 169 rem.12	**18.** 1567 rem.4		**19.** 1705 rem.26
20. 749	**21.** 189 rem.14		**22.** 14 638
23. 3600	**24.** 1331	**25.** 2128	

EXERCISE 54 (page 83)

Test 1

£	£	£	£
1. 81·25	**2.** 235·58	**3.** 1631·44$\frac{1}{2}$	**4.** 145·27$\frac{1}{2}$
5. 237·28$\frac{1}{2}$	**6.** 92·48	**7.** 237·12	**8.** 879·68
9. 1631·63$\frac{1}{2}$	**10.** 17·24		**11.** 20·91$\frac{1}{2}$ rem.5p
12. 18·98 rem.$\frac{1}{2}$p	**13.** 71·49 rem.15p		**14.** 53·50
15. 642·00	**16.** 1861·87$\frac{1}{2}$		

£

17. (a) 135·53   **18.** 46
    (b) 267·62   **19.** 49
    (c) 324·50$\frac{1}{2}$ **20.** 24
    (d) 127·02
    (e) 288·27
    (f) 225·76$\frac{1}{2}$
    (g) 206·57
    (h) 134·07
Total 854·67$\frac{1}{2}$

Test 2

£	£	£	£
1. 212·59	2. 427·01	3. 37·76½	4. 450·23½
5. 396·29	6. 187·72	7. 313·32	8. 1578·96
9. 2030·93	10. 3044·12½	11. 15·16½ rem.4p	
12. 15·15 rem.6p	13. 35·14 rem.8p	14. 23·69 rem.19p	
15. 217·80	16. 17·15	17. 24·50	18. 210·00

£
19. 3·96   20. 47
     6·63
    10·54
     3·99
     8·00
    ─────
    33·12

Test 3

£	£	£	£
1. 185·05	2. 515·70½	3. 371·18½	4. 8366·34
5. 118·97	6. 309·70½	7. 399·92½	8. 24·583
9. 779·10	10. 1479·85½	11. 4869·53½	
12. 16·06½ rem.2p	13. 12·23	14. 16·79½ rem.7p	
15. 35·57 rem.5p	16. 8·60	17. 180·36	
18. 35	19. 200 rem.25p		

£
20.   17·82
          1·16
          7·56
          5·98
          2·64
          4·41
         10·85
          2·16
          2·95
         ─────
Total   55·53

EXERCISE 55 (page 85)

Test 1

1. 254.35	2. 278.1745	3. 323.585	4. 12.57
5. 70.277	6. 190.227	7. 8.19	8. 33.32

9. 2.223   10. 4.2159   11. 0.1813   12. 90.047

13. 14.217   14. 0.903   15. 500   16. 0.086

17. 0.95   18. 0.933   19. 4.777   20. 3.571

21. $\frac{33}{40}$ 22. $4\frac{29}{40}$ 23. $13\frac{19}{20}$ 24. $27\frac{19}{200}$

25. 109.95   26. 117.665   27. 95.63   28. 0.612

29. 0.031 08   30. 0.366 68

Test 2

1. 205.365   2. 142.585   3. 184.501   4. 4.1879

5. 90.895   6. 9.75   7. 27.13   8. 268.215

9. 62.401   10. 13.95   11. 0.0999   12. 0.7812

13. 0.081 12   14. 0.002 751 15. 5.7112   16. 3.317

17. 86.190   18. 6.629   19. 10.205   20. 9.632

21. 32.727   22. 38.244   23. 0.733   24. 0.538

25. 0.916   26. 2.666   27. $2\frac{13}{20}$ 28. $3\frac{9}{200}$

29. $7\frac{17}{200}$ 30. 0.02436

Test 3

1. 555.045   2. 27.719   3. 46.9505   4. 96.414

5. 51.255   6. 2.09   7. 11.773   8. 17.951

9. 74.431   10. 13.63   11. 39.69   12. 0.7448

13. 2.392   14. 1.8405   15. 0.2163   16. 0.3618

17. 0.000 001 12 18. 19.619   19. 55.580   20. 2.134

21. 1.404   22. 320.754   23. 0.428   24. 4.666

25. 3.566   26. $\frac{51}{200}$ 27. $4\frac{129}{500}$ 28. $3\frac{113}{200}$

29. 46.399   30. 24.98

EXERCISE 56 (page 87)

Test 1

1. 138.1 cm   2. 2.666 m   3. 8.16 m

4. 2.408 tonnes   5. 2.503 litres   6. 3.6 kg

7. 2.551 litres   8. 106.257 km   9. 33.053 g

10. 278.635 litres 11. 3350 g   12. 0.5378 tonnes

13. 1.421 m   14. 11.34 kg   15. 11.844 tonnes

16. 5.625 litres   17. 5.688 m   18. 1128 dm

19. 38 080 m   20. 2.2 mm   21. 4.8 cm

22. 420 kg   23. 0.35 litres   24. 0.36 litres

25. 4.9 dm

Test 2

1. 1.638 kg
2. 4.322 kg
3. 552.288 m
4. 2.9491 km
5. 3.095 m
6. 2.175 litres
7. 2.819 tonnes
8. 25 364 000 mm
9. 33.565 m
10. 4.45 m
11. 3650 g
12. 0.788 55 tonnes
13. 1211.5 km
14. 23.75 tonnes
15. 1.728 m
16. 7.525 kg
17. 4.644 tonnes
18. 828 dm
19. 28 080 m
20. 0.1204 km
21. 4.7 mm
22. 0.041 tonnes
23. 0.041 litres
24. 0.49 litres
25. 470 kg

Test 3

1. 189.9 cm
2. 2.745 m
3. 1.36135 km
4. 3.473 litres
5. 10.78 kg
6. 2.598 litres
7. 1.812 tonnes
8. 254.58 m
9. 20.894 km
10. 25.705 g
11. 28.687 kg
12. 130.575 litres
13. 5.779 g
14. 8.255 km
15. 4789.5 kg
16. 24 450 m
17. 1.381 75 km
18. 7.866 m
19. 0.374 22 km
20. 32.625 kg
21. 5.44 g
22. 49 950 m
23. 390 kg
24. 390 m
25. 4.7 kg

EXERCISE 57 (page 89)

Test 1

1. $10\frac{23}{30}$
2. $9\frac{11}{40}$
3. $10\frac{23}{60}$
4. $13\frac{1}{2}$
5. $3\frac{7}{20}$
6. $1\frac{7}{12}$
7. $3\frac{47}{120}$
8. $7\frac{1}{2}$
9. $10\frac{1}{4}$
10. $7\frac{5}{16}$
11. $11\frac{11}{24}$
12. $4\frac{1}{3}$
13. 3
14. $2\frac{29}{32}$
15. $1\frac{1}{7}$
16. $\frac{9}{11}$
17. £224·25
18. £162
19. 5250 m
20. £5·18
21. $\frac{21}{59}$
22. $\frac{1}{4}$
23. $\frac{11}{18}$
24. 1246
25. £132·63
26. £57·71$\frac{1}{2}$
27. 9
28. $3\frac{1}{33}$
29. $6\frac{23}{24}$
30. $3\frac{7}{24}$

Test 2

1. $7\frac{2}{15}$
2. $10\frac{13}{20}$
3. $8\frac{9}{20}$
4. $11\frac{3}{8}$
5. $9\frac{9}{14}$
6. $1\frac{5}{12}$
7. $1\frac{41}{60}$
8. $2\frac{17}{60}$
9. 4
10. $6\frac{1}{2}$
11. 10
12. 14
13. $11\frac{31}{64}$
14. 10
15. $2\frac{1}{3}$
16. $2\frac{2}{5}$
17. £42
18. £80
19. 3850 kg

20. £40·31½ 21. £150　22. 3　　23. 6¾　　24. $\frac{29}{45}$
25. £54·98 26. 1516　27. 5730 kg　　　28. £149·59
29. $2\frac{47}{60}$　30. $6\frac{11}{15}$

Test 3

1. $7\frac{7}{30}$　　2. $7\frac{19}{30}$　　3. $8\frac{3}{7}$　　4. $7\frac{1}{24}$　　5. $14\frac{11}{45}$
6. $2\frac{7}{15}$　　7. $8\frac{5}{24}$　　8. $1\frac{31}{60}$　　9. 4　　10. 12
11. $24\frac{3}{4}$　12. $5\frac{1}{3}$　　13. 2　　14. $2\frac{17}{36}$　15. $1\frac{1}{5}$
16. £77　17. £280　18. £86·87 19. 7120 km 20. £60·77½
21. $4\frac{1}{2}$　22. $1\frac{4}{5}$　23. $\frac{67}{144}$　24. $1\frac{29}{66}$　25. £52·36
26. 5236　27. 4477½ 28. 1.575 litres　29. $8\frac{23}{30}$
30. $3\frac{1}{8}$

EXERCISE 58 (page 91)

Test 1

	Perimeter	Area		Perimeter	Area
1.	158 mm	1548 mm²	2.	15.2 m	13.63 m²
3.	31 m	59.16 m²	4.	28.8 m	46.55 m²
5.	10.1 m	5.85 m²	6.	15.9 m	15.2 m²
7.	31.3 m	61.09 m²	8.	44.7 m	120.15 m²
9.	119.4 m	762.2 m²	10.	21.05 km	26.95 km²

11. (a) 13.2 m²　　　　　　12. (a) 756 mm²
　　(b) 13.6 m²　　　　　　　　(b) 2646 mm²
　　(c) 12 m²　　　　　　　　　(c) 7.68 m²
　　(d) 8 m²　　　　　　　　　(d) 5.85 m²
　　(e) 8.25 m²　　　　　　　(e) 2.125 m²
　　(f) 15.96 m²　　　　　13. 48 mm
　　(g) 14.28 m²　　　　　14. £56·70
　　(h) 22.26 m²
　　(i) 18.99 m²
　　(j) £52·40
　　(k) £56·44
　　(l) £35·06
　　(m)£67·03

Test 2

	Perimeter	Area		Perimeter	Area
1.	154 mm	1372 mm²	2.	20.2 m	24.18 m²
3.	13.1 m	9.775 m²	4.	29.9 km	16.5 km²
5.	25.1 km	30.42 km²	6.	19.85 km	24.6 km²

7. (*a*) 494 m²   (*b*) 300 m²   (*c*) 25.48 m²
8. (*a*) £80·52   (*h*) 40 m²
   (*b*) £115·75   (*i*) 38 m²
   (*c*) £55·69   (*j*) 47.5 m²
   (*d*) £4·95   (*k*) 19 m
   (*e*) 36.5 m²   (*l*) 14.6 m
   (*f*) 37 m²   (*m*) 15.2 m
   (*g*) 29 m²   (*n*) 4.8 m
9. (*a*) 2.136 m²   (*d*) 6480 cm²
   (*b*) 4920 cm²   (*e*) 6650 cm²
   (*c*) 13.63 m²   (*f*) 3.6125 km²

EXERCISE 59 (page 94)

Test 1

1. (*a*) 2^3
   (*b*) $4^3 3^3$
   (*c*) $a^4 m^3$
   (*d*) $2^2 x^3 y^4$
   (*e*) $2^3 5^3 m^2 n^2 x$

2. (*a*) 16
   (*b*) 216
   (*c*) 1152
   (*d*) 4
   (*e*) 144
   (*f*) 108

3. (*a*) 4
   (*b*) 24
   (*c*) 6
   (*d*) 108
   (*e*) 0
   (*f*) 192
   (*g*) 0
   (*h*) 13 824

4. (*a*) x^9
   (*b*) y^5
   (*c*) $6x^7$
   (*d*) $3m^3$
   (*e*) $10x^6$
   (*f*) $4a^6$
   (*g*) $3ab^2 x$

5. (*a*) +25
   (*b*) −13
   (*c*) +14
   (*d*) +18

6. (*a*) −3
   (*b*) +3
   (*c*) +23
   (*d*) −62
   (*e*) $-9\frac{3}{4}$
   (*f*) −112
   (*g*) +6
   (*h*) −4.32

7. (*a*) +3
   (*b*) −2
   (*c*) $-\frac{1}{2}$
   (*d*) 1
   (*e*) $6x$
   (*f*) 16

8. (*a*) 6
   (*b*) 2
   (*c*) 18
   (*d*) $\frac{1}{4}$

   (*i*) +3
   (*j*) −12
   (*k*) +9
   (*l*) +4

1. (a) $3^4 5^2$
    (b) $2^2 x^3$
    (c) $x^3 y^3$
    (d) $3^3 4^2 x^3$
    (e) $2^2 3^2 5^2 x^3 y^2$

2. (a) 125
    (b) 576
    (c) 3456
    (d) 27
    (e) 36

3. (a) 9
    (b) 108
    (c) 0
    (d) 2880
    (e) 4608

4. (a) x^5
    (b) $12a^7$
    (c) $20a^3 b^5$
    (d) $7x^2$
    (e) $7x^2 y/3$
    (f) $6ab^2 x^2$

5. (a) +31
    (b) −29
    (c) +16
    (d) −7
    (e) +10
    (f) −3
    (g) $+14\frac{3}{4}$
    (h) +18
    (i) −7.7
    (j) 21.2

6. (a) +108
    (b) +168
    (c) $-15\frac{3}{4}$
    (d) $-3\frac{1}{3}$
    (e) −6
    (f) +4
    (g) +32

7. (a) $1\frac{1}{3}$
    (b) $\frac{1}{4}$
    (c) 1/108
    (d) −4
    (e) 28
    (f) 18
    (g) 54
    (h) 38

8. (a) 10
    (b) 10
    (c) −9
    (d) −10

9. 27 when $x = 3$
    16 when $x = 2$
    9 when $x = 1$
    6 when $x = 0$

Test 3

1. (a) 7^4
    (b) $6^3 4^4$
    (c) $x^4 y^3$
    (d) $3^4 a^4 x^3$

2. (a) 216
    (b) 2048
    (c) 100 000
    (d) 54
    (e) 108
    (f) $3\frac{3}{8}$

3. $16\frac{1}{3}$

4. $7a^3 bc^2$

5. (a) −12   (e) $-6\frac{1}{4}$
    (b) +7   (f) +12
    (c) +5   (g) −84
    (d) +5

6. 1152

7. (a) −12
    (b) $-1\frac{5}{29}$
    (c) $-2\frac{1}{2}$
    (d) 2

8. (a) $\frac{7}{25}$ (e) 3   9. 16 when $x = 3$
  (b) $\frac{1}{16}$ (f) $21x + 8$ 8 when $x = 2$
  (c) $\frac{3}{32}$ (g) 8   2 when $x = 1$
  (d) $\frac{1}{6}$ (h) -32 -2 when $x = 0$

EXERCISE 60 (page 98)

Test 1

1. (a) 20.72 m²   2. £130·65   3. £226·26
  (b) 25.3 m²   4. £26·49$\frac{1}{2}$ rem.7p   5. 548 rem.9p
6. 307 rem.10p   7.   £
           2·46
           9·36
           3·04$\frac{1}{2}$
           7·59$\frac{1}{2}$
           1·10$\frac{1}{2}$
           14·26$\frac{1}{2}$
           17·47$\frac{1}{2}$
           7·59$\frac{1}{2}$
     Total $\overline{62·90}$

8. 3.057 km   9. 3.132 litres   10. 5650 m
11. 5.977 m   12. 1.351 km   13. 8.1 dm   14. 0.372 litres
15. 212 m   16. $6\frac{1}{12}$ 17. $1\frac{1}{2}$ 18. $4\frac{11}{30}$
19. 2   20. £48·12$\frac{1}{2}$ 21. 43.2 m²   22. $\frac{4}{9}$
23. -2 24. 42 when $x = 4$ 25. 5.605 m²
           25 when $x = 3$
           12 when $x = 2$
           3 when $x = 1$

Test 2

1. 17 105   2. 5828   3. 96   4. £101·99$\frac{1}{2}$
5. £107·28$\frac{1}{2}$ 6. £304·55$\frac{1}{2}$ 7. £38·68$\frac{1}{2}$ rem.4$\frac{1}{2}$p
8. 116   9. (a)   31.819
           (b)   18.184
           (c)   25.456
           (d)   30.37
           (e)   44.654
           (f)   19.954
           (g)   25.918
           (h) $\underline{15.303}$
        Total 105.829

10. 4.97 **11.** 27.39 **12.** 270 **13.** 2.006 km
14. 3.196 litres **15.** 8.246 m **16.** 8.694 km **17.** 4100 m
18. 5.066 tonnes **19.** 0.41 kg **20.** 300 kg
21. (a) 118.8 m² **22.** 6085 mm² **23.** 29 when x is 3
    (b) 135 m²   16 when x is 2
    (c) 110.2 m²   7 when x is 1
    (d) 364 m²   2 when x is 0
            1 when x is -1
            4 when x is -2
24. $-2\frac{1}{3}$ **25.** (a) $+1\frac{5}{8}$ 11 when x is -3
       (b) -14
       (c) $+1\frac{7}{8}$

Test 3

1. 2.67 m **2.** 2.759 kg **3.** 4945 g **4.** 2.953 litre
5. 9.116 m **6.** 17.575 kg **7.** 23.084 tonnes
8. 18 190 kg **9.** 780 kg **10.** (a) 447.85 m²
            (b) 40.08 m²
            (c) 232.72 m²
            (d) 175.05 m²
            (e) 86.8 m
11. 16 263 **12.** 311 rem.12 **13.** 25 rem.12
14. £13·48 rem.3p **15.** £256·90 **16.** £853·77$\frac{1}{2}$
17. £136·25 **18.** £17·64 **19.** £164·12 **20.** £111·09
21. 252 mm **22.** 37 when x is 3
       15 when x is 2
       1 when x is 1
       -5 when x is 0
       -3 when x is -1
       7 when x is -2
       25 when x is -3